tredition®

tredition was established in 2006 by Sandra Latusseck and Soenke Schulz. Based in Hamburg, Germany, tredition offers publishing solutions to authors and publishing houses, combined with worldwide distribution of printed and digital book content. tredition is uniquely positioned to enable authors and publishing houses to create books on their own terms and without conventional manufacturing risks.

For more information please visit: www.tredition.com

TREDITION CLASSICS

This book is part of the TREDITION CLASSICS series. The creators of this series are united by passion for literature and driven by the intention of making all public domain books available in printed format again - worldwide. Most TREDITION CLASSICS titles have been out of print and off the bookstore shelves for decades. At tredition we believe that a great book never goes out of style and that its value is eternal. Several mostly non-profit literature projects provide content to tredition. To support their good work, tredition donates a portion of the proceeds from each sold copy. As a reader of a TREDITION CLASSICS book, you support our mission to save many of the amazing works of world literature from oblivion. See all available books at www.tredition.com.

 Project Gutenberg

The content for this book has been graciously provided by Project Gutenberg. Project Gutenberg is a non-profit organization founded by Michael Hart in 1971 at the University of Illinois. The mission of Project Gutenberg is simple: To encourage the creation and distribution of eBooks. Project Gutenberg is the first and largest collection of public domain eBooks.

The Brain and the Voice in Speech and Song

Frederick Walker Mott

Imprint

This book is part of TREDITION CLASSICS

Author: Frederick Walker Mott
Cover design: Buchgut, Berlin – Germany

Publisher: tredition GmbH, Hamburg - Germany
ISBN: 978-3-8424-3463-9

www.tredition.com
www.tredition.de

Copyright:
The content of this book is sourced from the public domain.

The intention of the TREDITION CLASSICS series is to make world literature in the public domain available in printed format. Literary enthusiasts and organizations, such as Project Gutenberg, worldwide have scanned and digitally edited the original texts. tredition has subsequently formatted and redesigned the content into a modern reading layout. Therefore, we cannot guarantee the exact reproduction of the original format of a particular historic edition. Please also note that no modifications have been made to the spelling, therefore it may differ from the orthography used today.

THE BRAIN AND THE VOICE IN SPEECH AND SONG

BY F.W. MOTT, F.R.S., M.D., F.R.C.P.

1910

PREFACE

The contents of this little book formed the subject of three lectures delivered at the Royal Institution "On the Mechanism of the Human Voice" and three London University lectures at King's College on "The Brain in relation to Speech and Song." I have endeavoured to place this subject before my readers in as simple language as scientific accuracy and requirements permit. Where I have been obliged to use technical anatomical and physiological terms I have either explained their meaning in the text, aided by diagrams and figures, or I have given in brackets the English equivalents of the terms used.

I trust my attempt to give a sketch of the mechanism of the human voice, and how it is produced in speech and song, may prove of interest to the general public, and I even hope that teachers of voice production may find some of the pages dealing with the brain mechanism not unworthy of their attention.

F.W. MOTT

LONDON

July, 1910

CONTENTS

THEORIES ON THE ORIGIN OF SPEECH

THE VOCAL INSTRUMENT:

THREE QUALITIES OF MUSICAL SOUNDS, LOUDNESS, PITCH AND TIMBRE

THE VOCAL INSTRUMENT AND ITS THREE PARTS

(1) THE BELLOWS AND ITS STRUCTURE: VOLUNTARY CONTROL OF BREATH

(2) THE REED CONTAINED IN THE VOICE-BOX OR LARYNX: ITS STRUCTURE AND ACTION

(3) THE RESONATOR AND ARTICULATOR, ITS STRUCTURE AND ACTION IN SONG AND SPEECH

PATHOLOGICAL DEGENERATIVE CHANGES PRODUCING SPEECH DEFECTS AND WHAT THEY TEACH

THE CEREBRAL MECHANISM OF SPEECH AND SONG

SPEECH AND RIGHT-HANDEDNESS

LOCALISATION OF SPEECH CENTRES IN THE BRAIN

THE PRIMARY SITE OF REVIVAL OF WORDS IN SILENT THOUGHT

CASE OF DEAFNESS ARISING FROM DESTRUCTION OF THE AUDITORY CENTRES IN THE BRAIN CAUSING LOSS OF SPEECH

THE PRIMARY REVIVAL OF SOME SENSATIONS IN THE BRAIN

PSYCHIC MECHANISM OF THE VOICE

ILLUSTRATIONS

FIG.

1. The thoracic cage and its contents

2. The diaphragm and its attachments

3. Diagram illustrating changes of the chest and abdomen in breathing

4. Diagram of the cartilages of the voice-box or larynx with vocal cords

5. Front view of the larynx with muscles

6. Back view of the larynx with muscles

7. Diagram to illustrate movements of cartilages in breathing and phonation

8. Section through larynx and windpipe, showing muscles and vocal cords

9. The laryngoscope and its use

10. The glottis in breathing, whispering, and vocalisation

11. The vocal cords in singing, after French

12. Vertical section through the head and neck to show the larynx and resonator

13. Diagram (after Aikin) of the resonator in the production of the vowel sounds

14. Kvnig's flame manometer

15. Diagram of a neurone

16. Left hemisphere, showing cerebral localisation

17. Diagram to illustrate cerebral mechanism of speech, after Bastian

18. The course of innervation currents in phonation

THE BRAIN AND THE VOICE IN SPEECH AND SONG

In the following pages on the Relation of the Brain to the mechanism of the Voice in Speech and Song, I intend, as far as possible, to explain the mechanism of the instrument, and what I know regarding the cerebral mechanism by which the instrument is played upon in the production of the singing voice and articulate speech. Before, however, passing to consider in detail the instrument, I will briefly direct your attention to some facts and theories regarding the origin of speech.

THEORIES ON THE ORIGIN OF SPEECH

The evolutionary theory is thus propounded by Romanes in his "Mental Evolution in Man," pp. 377-399: "Starting from the highly intelligent and social species of anthropoid ape as pictured by Darwin, we can imagine that this animal was accustomed to use its voice freely for the expression of the emotions, uttering danger signals, and singing. Possibly it may also have been sufficiently intelligent to use a few imitative sounds; and certainly sooner or later the receptual life of this social animal must have advanced far enough to have become comparable with that of an infant of about two years of age. That is to say, this animal, although not yet having begun to use articulate signs, must have advanced far enough in the conventional use of natural signs (a sign with a natural origin in tone and gesture, whether spontaneously or intentionally imitative) to have admitted of a totally free exchange of receptual ideas, such as would be concerned in animal wants and even, perhaps, in the simplest forms of co-operative action. Next I think it probable that the advance of receptual intelligence which would have been occasioned by this advance in sign-making would in turn have led to a development of the latter — the two thus acting and reacting on each other until the language of tone and gesture became gradually raised to the level of imperfect pantomime, as in children before they begin to use words. At this stage, however, or even before it, I think very probably vowel sounds must have been employed in tone language, if not also a few consonants. Eventually the action and reaction of receptual intelligence and conventional sign-making must have ended in so far developing the former as to have admitted of the breaking up (or articulation) of vocal sounds, as the only direction in which any improvement in vocal sign-making was possible." Romanes continues his sketch by referring to the probability that this important stage in the development of speech was greatly assisted by the already existing habit of articulating musical notes, supposing our progenitors to have resembled the gibbons or the chimpanzees in this respect. Darwin in his great work on the "Expression of the Emotions" points to the fact that the gibbon, the most erect and active of the anthropoid apes, is able to sing an octave in half-tones, and it is interesting to note that Dubois considers his

Pithecanthropus Erectus is on the same stem as the gibbon. But it has lately been shown that some animals much lower in the scale than monkeys, namely, rodents, are able to produce correct musical tones. Therefore the argument loses force that the progenitors of man probably uttered musical sounds before they had acquired the power of articulate speech, and that consequently, when the voice is used under any strong emotion, it tends to assume through the principle of association a musical character. The work of anthropologists and linguists, especially the former, supports the progressive-evolution theory, which, briefly stated, is — that articulate language is the result of an elaboration in the long procession of ages in which there occurred three stages — the cry, vocalisation, and articulation. The cry is the primordial, pure animal language; it is a simple vocal aspiration without articulation; it is either a reflex expressing needs and emotions, or at a higher stage intentional (to call, warn, menace, etc.). Vocalisation (emission of vowels) is a natural production of the vocal instrument, and does not in itself contain the essential elements of speech. Many animals are capable of vocalisation, and in the child the utterance of vowel sounds is the next stage after the cry.

The conditions necessary to the existence of speech arose with articulation, and it is intelligence that has converted the vocal instrument into the speaking instrument. For whereas correct intonation depends upon the innate musical ear, which is able to control and regulate the tensions of the minute muscles acting upon the vocal cords, it is intelligence which alters and changes the form of the resonator by means of movement of the lips, tongue, and jaw in the production of articulate speech. The simple musical instrument in the production of phonation is bilaterally represented in the brain, but as a speaking instrument it is unilaterally represented in right-handed individuals in the left hemisphere and in left-handed individuals in the right hemisphere. The reason for this we shall consider later; but the fact supports Darwin's hypothesis.

Another hypothesis which was brought forward by Grieger and supported by some authors is summarised by Ribot as follows: "Words are an imitation of the movements of the mouth. The predominant sense in man is that of sight; man is pre-eminently visual. Prior to the acquisition of speech he communicated with his fellows by the aid of gestures and movement of the mouth and face; he appealed to their eyes. Their facial 'grimaces,' fulfilled and elucidated by gesture, became signs for others; they fixed their attention upon them. When articulate sounds came into being, these lent themselves to a more or less conventional language by reason of their acquired importance." For support of this hypothesis the case of non-educated deaf-mutes is cited. They invent articulate sounds which they cannot hear and use them to designate certain things. Moreover, they employ gesture language—a language which is universally understood.

Another theory of the origin of the speaking voice is that speech is an instinct not evolved, but breaking forth spontaneously in man; but even if this be so, it was originally so inadequate and weak that it required support from the gesture language to become intelligible. This mixed language still survives among some of the inferior races of men. Miss Kingsley and Tylor have pointed out that tribes in Africa have to gather round the camp fires at night in order to converse, because their vocabulary is so incomplete that without being reinforced by gesture and pantomime they would be unable to communicate with one another. Gesture is indispensable for giving precision to vocal sounds in many languages, e.g. those of the Tasmanians, Greenlanders, savage tribes of Brazil, and Grebos of Western Africa. In other cases speech is associated with inarticulate sounds. These sounds have been compared to clicking and clapping, and according to Sayce, these clickings and clappings survive as though to show us how man when deprived of speech can fix and transmit his thoughts by certain sounds. These mixed states represent articulate speech in its primordial state; they represent the stage of transition from pure pantomime to articulate speech.

It seems, then, that originally man had two languages at his disposal which he used simultaneously or interchangeably. They supported each other in the intercommunication of ideas, but speech has triumphed because of its greater practical utility. The language of gesture is disadvantageous for the following reasons: (1) it monopolises the use of the hands; (2) it has the disadvantage that it does not carry any distance; (3) it is useless in the dark; (4) it is vague in character; (5) it is imitative in nature and permits only of the intercommunication of ideas based upon concrete images.

Speech, on the other hand, is transmitted in the dark and with objects intervening; moreover, distance affects its transmission much less. The images of auditory and visual symbols in the growth of speech replace in our minds concrete images and they permit of abstract thought. It is dependent primarily upon the ear, an organ of exquisite feeling, whose sensations are infinite in number and in kind. This sensory receptor with its cerebral perceptor has in the long process of time, aided by vision, under the influence of natural laws of the survival of the fittest, educated and developed an instrument of simple construction (primarily adapted only for the vegetative functions of life and simple vocalisation) into that wonderful instrument the human voice; but by that development, borrowing the words of Huxley, "man has slowly accumulated and organised the experience which is almost wholly lost with the cessation of every individual life in other animals; so that now he stands raised as upon a mountain-top, far above the level of his humble fellows, and transfigured from his grosser nature by reflecting here and there a ray from the infinite source of truth." Thought in all the higher mental processes could not be carried on at all without the aid of language.

Written language probably originated in an analytical process analogous to the language of gesture. Like that, it: (1) isolates terms; (2) arranges them in a certain order; (3) translates thoughts in a crude and somewhat vague form. A curious example of this may be found in Max M|ller's "Chips from a German Workshop," XIV.: "The aborigines of the Caroline Islands sent a letter to a Spanish captain as follows: A man with extended arms, sign of greeting; below to the left, the objects they have to barter—five big shells, seven little ones, three others of different forms; to the right, drawing of the objects they wanted in exchange—three large fish-hooks, four small ones, two axes, two pieces of iron."

Language of graphic signs and spoken language have progressed together, and simultaneously supported each other in the development of the higher mental faculties that differentiate the savage from the brute and the civilised human being from the savage. In spoken language, at any rate, it is not the vocal instrument that has been changed, but the organ of mind with its innate and invisible molecular potentialities, the result of racial and ancestral experiences in past ages. Completely developed languages when studied from the point of view of their evolution are stamped with the print of an unconscious labour that has been fashioning them for centuries. A little consideration and reflection upon words which have been coined in our own time shows that language offers an abstract and brief chronicle of social psychology.

Articulate language has converted the vocal instrument into the chief agent of the will, but the brain in the process of time has developed by the movements of the lips, tongue, jaw, and soft palate a kinfsthetic9 sense of articulate speech, which has been integrated and associated in the mind with rhythmical modulated sounds conveyed to the brain by the auditory nerves. There has thus been a reciprocal simultaneity in the development of these two senses by which the mental ideas of spoken words are memorised and recalled. Had man been limited to articulate speech he could not have made the immense progress he has made in the development of complex mental processes, for language, by using written verbal symbols, has allowed, not merely the transmission of thought from one individual to another, but the thoughts of the world, past and present, are in a certain measure at the disposal of every individual. With this introduction to the subject I will pass on to give a detailed description of the instrument of the voice.

[Footnote 1: Sense of movement.]

THE VOCAL INSTRUMENT

A distinction is generally made in physics between sound and noise. Noise affects our tympanic membrane as an irregular succession of shocks and we are conscious of a jarring of the auditory apparatus; whereas a musical sound is smooth and pleasant because the tympanic membrane is thrown into successive periodic vibrations to which the auditory receptor (sense organ of hearing) has been attuned. To produce musical sounds, a body must vibrate with the regularity of a pendulum, but it must be capable of imparting sharper or quicker shocks to the air than the pendulum. All musical sounds, however they are produced and by whatever means they are propagated, may be distinguished by three different qualities:

(1) Loudness, (2) Pitch, (3) Quality, timbre or klang, as the Germans call it.

Loudness depends upon the amount of energy expended in producing the sound. If I rub a tuning-fork with a well-rosined bow, I set it in vibration by the resistance offered to the rosined hair; and if while it is vibrating I again apply the bow, thus expending more energy, the note produced is louder. Repeating the action several times, the width of excursion of the prongs of the tuning-fork is increased. This I can demonstrate, not merely by the loudness of the sound which can be heard, but by sight; for if a small mirror be fixed on one of the prongs and a beam of light be cast upon the mirror, the light being again reflected on to the screen, you will see the spot of light dance up and down, and the more energetically the tuning-fork is bowed the greater is the amplitude of the oscillation of the spot of light. The duration of the time occupied is the same in traversing a longer as in traversing a shorter space, as is the case of the swinging pendulum. The vibrating prongs of the tuning-fork throw the air into vibrations which are conveyed to the ear and produce the sensation of sound. The duration of time occupied in the vibrations of the tuning-fork is therefore independent of the space passed over. The greater or less energy expended does not influence the duration of time occupied by the vibration; it only influences the amplitude of the vibration.

The second quality of musical sounds is the pitch, and the pitch depends upon the number of vibrations that a sounding body makes in each second of time. The most unmusical ear can distinguish a high note from a low one, even when the interval is not great. Low notes are characterised by a relatively small number of vibrations, and as the pitch rises so the number of vibrations increase. This can be proved in many ways. Take, for example, two tuning-forks of different size: the shorter produces a considerably higher pitched note than the longer one. If a mirror be attached to one of the prongs of each fork, and a beam of light be cast upon each mirror successively and then reflected in a revolving mirror, the oscillating spot of light is converted into a series of waves; and if the waves obtained by reflecting the light from the mirror of the smaller one be counted and compared with those reflected from the mirror attached to the larger fork, it will be found that the number of waves reflected from the smaller fork is proportionally to the difference in the pitch more numerous than the waves reflected from the larger. The air is thrown into corresponding periodic vibrations according to the rate of vibration of the sound-producing body.

Thirdly, the quality, timbre, or klang depends upon the overtones, in respect to which I could cite many experiments to prove that whenever a body vibrates, other bodies near it may be set in vibration, but only on condition that such bodies shall be capable themselves of producing the same note. A number of different forms of resonators can be used to illustrate this law; a law indeed which is of the greatest importance in connection with the mechanism of the human voice. Although notes are of the same loudness and pitch when played on different instruments or spoken or sung by different individuals, yet even a person with no ear for music can easily detect a difference in the quality of the sound and is able to recognise the nature of the instrument or the timbre of the voice. This difference in the timbre is due to harmonics or overtones. Could we but see the sonorous waves in the air during the transmission of the sound of a voice, we should see stamped on it the conditions of motion upon which its characteristic qualities depended; which is due to the fact that every vocal sound whose vibrations have a complex form can be decomposed into a series of simple notes all belonging to the harmonic series. These harmonics or overtones will be considered later when dealing with the timbre or quality of the human voice.

The vocal instrument is unlike any other musical instrument; it most nearly approaches a reed instrument. The clarionet and the oboe are examples of reed instruments, in which the reed does not alter but by means of stops the length of the column of air in the resonating pipe varies and determines the pitch of the fundamental note. The organ-pipe with the vibrating tongue of metal serving as the reed is perhaps the nearest approach to the vocal organ; but here again it is the length of the pipe which determines the pitch of the note.

The vocal instrument may be said to consist of three parts: (1) the bellows; (2) the membranous reed contained in the larynx, which by the actions of groups of muscles can be altered in tension and thus variation in pitch determined; (3) the resonator, which consists of the mouth, the throat, the larynx, the nose, and air sinuses contained in the bones of the skull, also the windpipe, the bronchial tubes, and the lungs. The main and important part of the resonator, however, is situated above the glottis (the opening between the vocal cords, *vide* fig. 6), and it is capable of only slight variations in length and of many and important variations in form. In the production of musical sounds its chief influence is upon the quality of the overtones and therefore upon the timbre of the voice; moreover, the movable structures of the resonator, the lower jaw, the lips, the tongue, the soft palate, can, by changing the form of the resonator, not only impress upon the sound waves particular overtones as they issue from the mouth, but simultaneously can effect the combination of vowels and consonants with the formation of syllables, the combination of syllables with the formation of words, and the combination of words with the formation of articulate language. The reed portion of the instrument acting alone can only express emotional feeling; the resonator, the effector of articulate speech, is the instrument of intelligence, will, and feeling. It must not, however, be thought that the vocal instrument consists of two separately usable parts, for phonation (except in the whispered voice) always accompanies articulation.

In speech, and more especially in singing, there is an art of breathing. Ordinary inspiration and expiration necessary for the oxygenation of the blood is performed automatically and unconsciously. But in singing the respiratory apparatus is used like the bellows of a musical instrument, and it is controlled and directed by the will; the art of breathing properly is fundamental for the proper production of the singing voice and the speaking voice of the orator. It is necessary always to maintain in the lungs, which act as the bellows, a sufficient reserve of air to finish a phrase; therefore when the opportunity arises it is desirable to take in as much air as possible through the nostrils, and without any apparent effort; the expenditure of the air in the lungs must be controlled and regulated by the power of the will in such a manner as to produce efficiency in loudness with economy of expenditure. It must be remembered, moreover, that mere loudness of sound does not necessarily imply carrying power of the voice, either when speaking or singing. Carrying power, as we shall see later, depends as much upon the proper use of the resonator as upon the force of expulsion of the air by the bellows. Again, a soft note, especially an aspirate, owing to the vocal chink being widely opened, may be the cause of an expenditure of a larger amount of air than a loud-sounding note. Observations upon anencephalous monsters (infants born without the great brain) show that breathing and crying can occur without the cerebral hemispheres; moreover, Goltz's dog, in which all the brain had been removed except the stem and base, was able to bark, growl, and snarl, indicating that the primitive function of the vocal instrument can be performed by the lower centres of the brain situated in the medulla oblongata. But the animal growled and barked when the attendant, who fed it daily, approached to give it food, which was a clear indication that the bark and growl had lost both its affective and cognitive significance; it was, indeed, a purely automatic reflex action. It was dependent upon a stimulus arousing an excitation in an instinctive automatic nervous mechanism in the medulla oblongata and spinal cord presiding over synergic groups of muscles habitually brought into action for this simplest form of vocalisation, connected with the primitive emotion of anger.

I will now consider at greater length each part of the vocal instrument.

I. THE BELLOWS

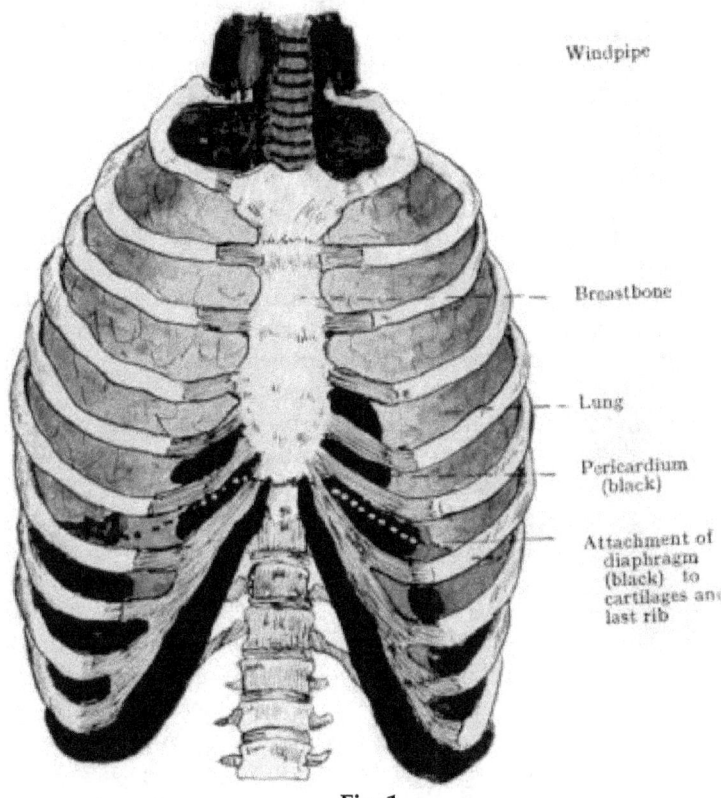

Fig. 1

FIG. 1.—Front view of the thorax showing the breastbone, to which on either side are attached the (shaded) rib cartilages. The remainder of the thoracic cage is formed by the ribs attached behind to the spine, which is only seen below. The lungs are represented filling the chest cavity, except a little to the left of the breastbone, below where the pericardium is shown (black). It can be seen that the ribs slope forwards and downwards, and that they increase in length from above downwards, so that if elevated by the muscles attached to them, they will tend to push forward the elastic cartilages and breastbone and so increase the antero-posterior diameter of the chest; moreover, the ribs being elastic will tend to give a little at the angle, and so the lateral diameter of the chest will be increased.

The bellows consists of the lungs enclosed in the movable thorax. The latter may be likened to a cage; it is formed by the spine behind and the ribs, which are attached by cartilages to the breastbone (sternum) in front (*vide* fig. 1). The ribs and cartilages, as the diagram shows, form a series of hoops which increase in length from above downwards; moreover, they slope obliquely downwards and inwards (*vide* fig. 2). The ribs are jointed behind to the vertebrae in such a way that muscles attached to them can, by shortening, elevate them; the effect is that the longer ribs are raised, and pushing forward the breastbone and cartilages, the thoracic cage enlarges from before back; but being elastic, the hoops will give a little and cause some expansion from side to side; moreover, when the ribs are raised, each one is rotated on its axis in such a way that the lower border tends towards eversion; the total effect of this rotation is a lateral expansion of the whole thorax. Between the ribs and the cartilages the space is filled by the intercostal muscles (*vide* fig. 2), the action of which, in conjunction with other muscles, is to elevate the ribs. It is, however, unnecessary to enter into anatomical details, and describe all those muscles which elevate and rotate the ribs, and thereby cause enlargement of the thorax in its antero-posterior and lateral diameters. There is, however, one muscle which forms the floor of the thoracic cage called the diaphragm that requires more than a passing notice (*vide* fig. 2), inasmuch as it is the most effective agent in the expansion of the chest. It consists of a central tendinous portion, above which lies the heart, contained in its bag or pericardium; on either side attached to the central tendon on the one hand and to the spine behind, to the last rib laterally, and to the cartilages of the lowest six ribs anteriorly, is a sheet of muscle fibres which form on either side of the chest a dome-like partition between the lungs and the abdominal cavity (*vide* fig. 2). The phrenic nerve arises from the spinal cord in the upper cervical region and descends through the neck and chest to the diaphragm; it is therefore a special nerve of respiration. There are two—one on each side supplying the two sheets of muscle fibres. When innervation currents flow down these nerves the two muscular halves of the diaphragm contract, and the floor of the chest on either side descends; thus the vertical diameter increases. Now the elastic lungs are covered with a smooth pleura which is reflected from them on to the inner side of the wall of the thorax, leaving no space between; consequently when the chest expands in all three directions the elastic lungs expand correspondingly. But when either voluntarily or automatically the nerve currents that cause contraction of the muscles of expansion cease, the elastic structures of the lungs and thorax, including the muscles, recoil, the diaphragm ascends, and the ribs by the force of gravity tend to fall into the position of rest. During expansion of the chest a negative pressure is established in the air passages and air flows into them from without. In contraction of the chest there is a positive pressure in the air passages, and air is expelled; in normal quiet breathing an ebb and flow of air takes place rhythmically and subconsciously; thus in the ordinary speaking of conversation we do not require to exercise any voluntary effort in controlling the breathing, but the orator and more especially the singer uses his knowledge and experience in the voluntary control of his breath, and he is thus enabled to use his vocal instrument in the most effective manner.

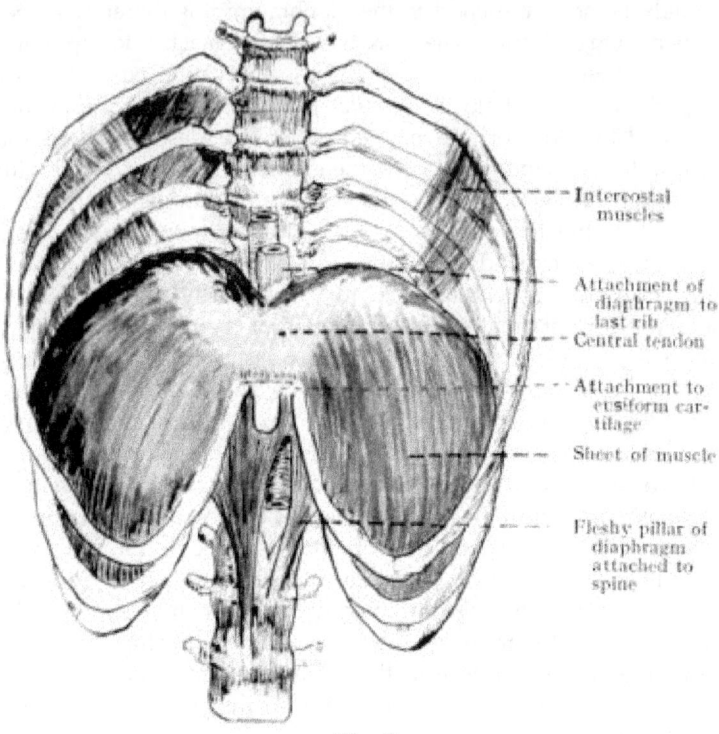

Fig. 2
Adapted from Quain's "Anatomy" by permission of Messrs. Longmans, Green & Co.

FIG. 2.—Diagram modified from Quain's "Anatomy" to show the attachment of the diaphragm by fleshy pillars to the spinal column, to the rib cartilages, and lower end of the breastbone and last rib. The muscular fibres, intercostals, and elevators of the ribs are seen, and it will be observed that their action would be to rotate and elevate the ribs. The dome-like shape of the diaphragm is seen, and it can be easily understood that if the central tendon is fixed and the sheet of muscle fibres on either side contracts, the floor of the chest on either side will flatten, allowing the lungs to expand vertically. The joints of the ribs with the spine can be seen, and the slope of the surface of the ribs is shown, so that when elevation and rotation occur the chest will be increased in diameter laterally.

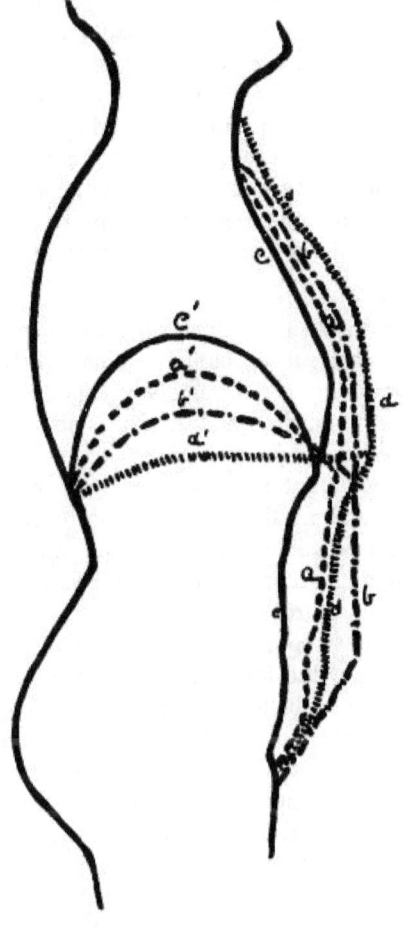

Fig. 3

FIG 3.—Diagram after Barth to illustrate the changes in the diaphragm, the chest, and abdomen in ordinary inspiration b-b', and expiration a-a', and in voluntary inspiration d-d' and expiration c-c', for vocalisation In normal breathing the position of the chest and abdomen in inspiration and expiration is represented respectively by the lines b and a; the position of the diaphragm is represented by b' and a'. In breathing for vocalisation the position of the chest and abdomen is represented by the lines d and e, and the diaphragm by d' and c'; it will be observed that in voluntary costal breathing d-d the expansion of the chest is much greater and also the diaphragm d' sinks deeper, but by the contraction of the abdominal muscles the protrusion of the belly wall d is much less than in normal breathing b.

A glance at the diagram (fig. 3) shows the changes in the shape of the thorax in normal subconscious automatic breathing, and the changes in the voluntary conscious breathing of vocalisation. It will be observed that there are marked differences: when voluntary control is exercised, the expansion of the chest is greater in all directions; moreover, by voluntary conscious effort the contraction of the chest is much greater in all directions; the result is that a larger amount of air can be taken into the bellows and a larger amount expelled. The mind can therefore bring into play at will more muscular forces, and so control and regulate those forces as to produce infinite variations in the pressure of the air in the sound-pipe of the vocal instrument. But the forces which tend to contract the chest and drive the air out of the lungs would be ineffective if there were not simultaneously the power of closing the sound-pipe; this we shall see is accomplished by the synergic action of the muscles which make tense and approximate the vocal cords. Although the elastic recoil of the lungs and the structure of the expanded thorax is the main force employed in normal breathing and to some extent in vocalisation (for it keeps up a constant steady pressure), the mind, by exercising control over the continuance of elevation of the ribs and contraction of the abdominal muscles, regulates the force of the expiratory blast of air so as to employ the bellows most efficiently in vocalisation. Not only does the contraction of the abdominal muscles permit of control over the expulsion of the air, but by fixing the cartilages of the lowest six ribs it prevents the diaphragm drawing them upwards and *inwards* (*vide* fig. 2). The greatest expansion is just above the waistband (*vide* fig. 3). We are not conscious of the contraction of the diaphragm; we are conscious of the position of the walls of the chest and abdomen; the messages the mind receives relating to the amount of air in the bellows at our disposal come from sensations derived from the structures forming the wall of the chest and abdomen, viz. the position of the ribs, their degree of elevation and forward protrusion combined with the feeling that the ribs are falling back into the position of rest; besides there is the feeling that the abdominal muscles can contract no more — a feeling which should never be allowed to arise before we become conscious of the necessity of replenishing the supply of air. This should be effected by quickly drawing in air through the nostrils without apparent effort and to as full extent as opportunity offers between the phrases. By intelligence and perseverance the guiding sense which informs the singer of the amount of air at his disposal, and when and how it should be replenished and voluntarily used, is of fundamental importance to good vocalisation. Collar-bone breathing is deprecated by some authorities, but I see no reason why the apices of the lungs should not be expanded, and seeing the frequency with which tubercle occurs in this region, it might by improving the circulation and nutrition be even beneficial. The proper mode of breathing comes almost natural to some individuals; to others it requires patient cultivation under a teacher who understands the art of singing and the importance of the correct methods of breathing.

The more powerfully the abdominal muscles contract the laxer must become the diaphragm muscle; and by the law of the reciprocal innervation of antagonistic muscles it is probable that with the augmented innervation currents to the expiratory centre of the medulla there is a corresponding inhibition of the innervation currents to the inspiratory centre (*vide* fig. 18, page 101). These centres in the medulla preside over the centres in the spinal cord which are in direct relation to the inspiratory and expiratory muscles. It is, however, probable that there is a direct relation between the brain and the spinal nerve centres which control the costal and abdominal muscles independently of the respiratory centres of the medulla oblongata (*vide* fig. 18). The best method of breathing is that which is most natural; there should not be a protruded abdomen on the one hand, nor an unduly inflated chest on the other hand; the maximum expansion should involve the lower part of the chest and the uppermost part of the abdomen on a level of an inch or more below the tip of the breastbone; the expansion of the ribs should be maintained as long as possible. In short phrases the movement may be limited to an ascent of the diaphragm, over which we have not the same control as we have of the elevation of the ribs; but it is better to reserve the costal air, over which we have more voluntary control, for maintaining a continuous pressure and for varying the pressure.

II. THE REED

I will now pass on to the consideration of the voice-box, or larynx, containing the reed portion of the vocal instrument.

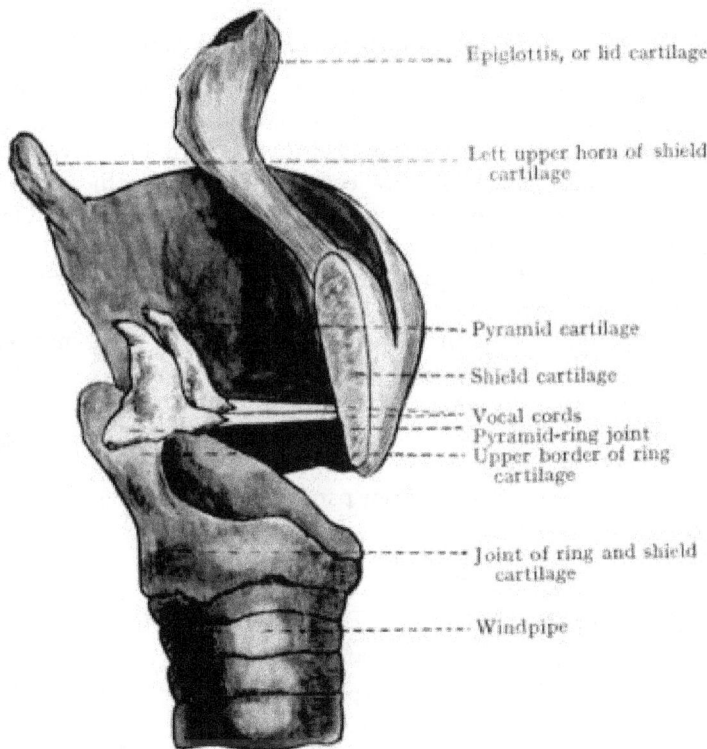

Fig. 4
from Behnke's 'Mechanism of the Human Voice'

FIG. 4.—The cartilages of the larynx or voice-box. A large portion of the shield cartilage on the right side has been cut away, in order to show the two pyramid cartilages; these are seen jointed by their bases with the ring cartilage; anteriorly are seen the two vocal processes which give attachment to the two vocal cords (white ligaments), which extend across the voice-box to be inserted in front in the angle of the shield cartilage. Groups of muscles pull upon these cartilages in such a manner as to increase, or diminish, the chink between the vocal cord in ordinary inspiration and expiration; in phonation a group of muscles approximate the cords, while another muscle makes them tense.

The Larynx.—The larynx is situated at the top of the sound-pipe (trachea or windpipe), and consists of a framework of cartilages articulated or jointed with one another so as to permit of movement (*vide* fig. 4). The cartilages are called by names which indicate their form and shape: (1) shield or thyroid, (2) the ring or cricoid, and (3) a pair of pyramidal or arytenoid cartilages. Besides these there is the epiglottis, which from its situation above the glottis acts more or less as a lid. The shield cartilage is attached by ligaments and muscles to the bone (hyoid) in the root of the tongue, a pair of muscles also connect this cartilage with the sternum or breastbone. The ring cartilage is attached to the windpipe by its lower border; by its upper border in front it is connected with the inner surface of the shield cartilage by a ligament; it is also jointed on either side with the shield cartilage. The posterior part of the ring cartilage is much wider than the anterior portion, and seated upon its upper and posterior rim and articulated with it by separate joints are the two pyramidal cartilages (*vide* fig. 4). The two vocal cords as shown in the diagram are attached to the shield cartilage in front, their attachments being close together; posteriorly they are attached to the pyramidal cartilages. It is necessary, however, to describe a little more fully these attachments. Extending forwards from the base of the pyramids are processes termed the "vocal processes," and these processes give attachment to the elastic fibres of which the vocal cords mainly consist. There are certain groups of muscles which by their attachment to the cartilages of the larynx and their action on the joints are able to separate the vocal cords or approximate them; these are termed respectively abductor and adductor muscles (figs. 5 and 6). In normal respiration the posterior ring-pyramidal muscles contract synergically with the muscles of inspiration and by separating the vocal cords open wide the glottis, whereby there is a free entrance of air to the windpipe; during expiration this muscle ceases to contract and the aperture of the glottis becomes narrower (*vide* fig. 10). But when the pressure is required to be raised in the air passages, as in the simple reflex act of coughing or in vocalisation, the glottis must be closed by approximation of the vocal cords, and this is effected by a group of muscles termed the adductors, which pull on the pyramid cartilages in such a way that the vocal processes are drawn towards one another in the manner shown in fig. 7. Besides the abductor and adductor groups of muscles, there is a muscle which acts in conjunction with the adductor group, and by its attachments to the shield cartilage above and the ring cartilage below makes tense the vocal cords (*vide* fig. 5); it is of interest to note that this muscle has a separate nerve supply to that of the abductor and adductor muscles.

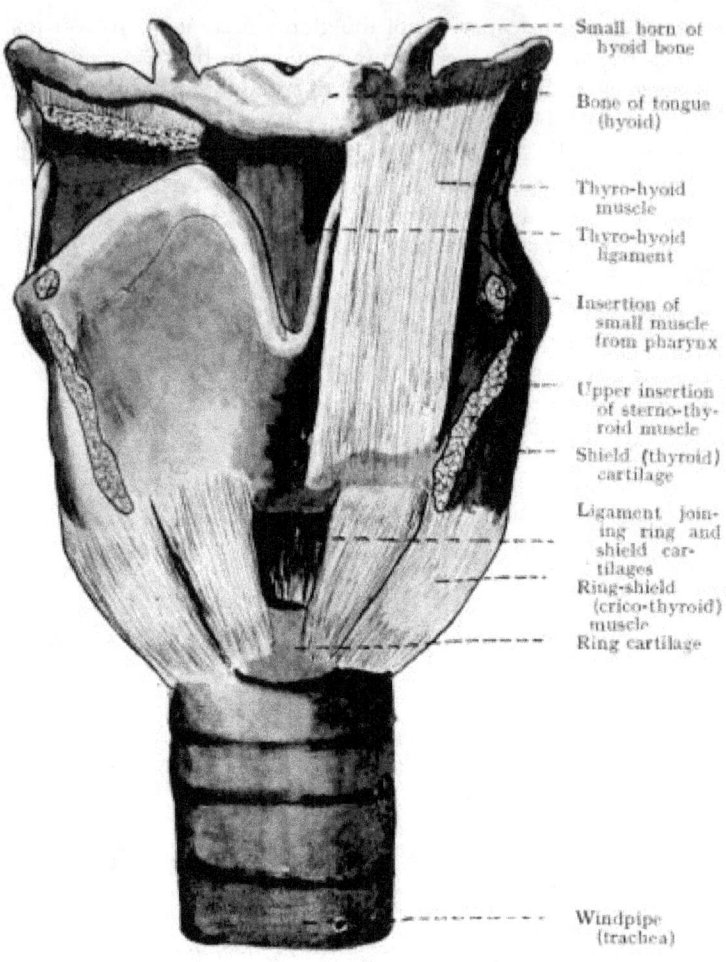

Fig. 5
Diagram after Testut (modified), showing the larynx from the front.

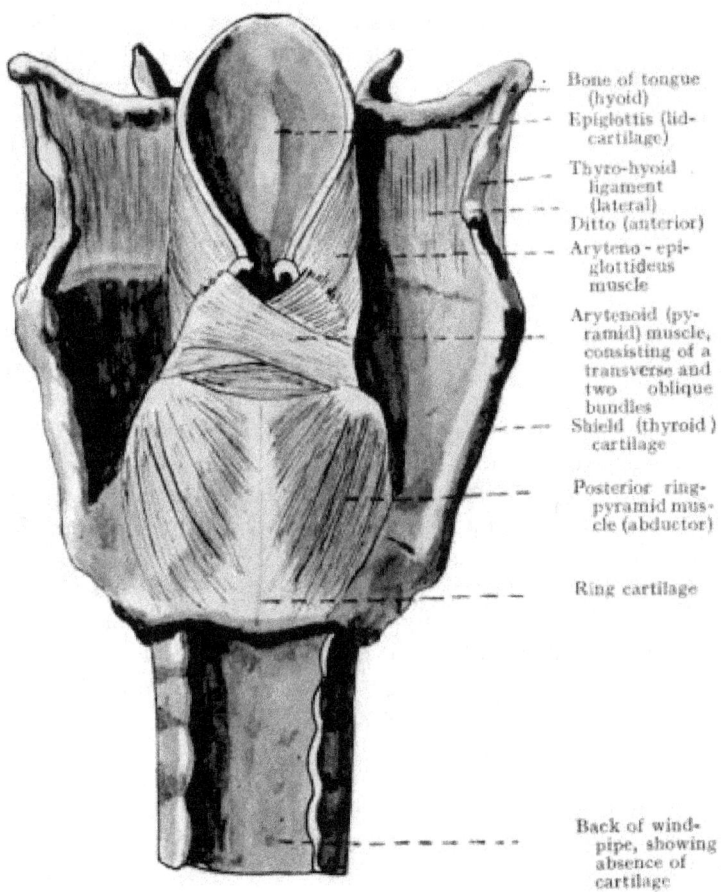

Fig. 6
Diagram after Testut (modified), showing the posterior view of the larynx with the muscles.

On the top of the pyramid cartilages, in the folds of mucous membrane which cover the whole inside of the larynx are two little pieces of yellow elastic cartilage; and in the folds of mucous membrane uniting these cartilages with the leaf-like lid cartilage (epiglottis) is a thin sheet of muscle fibres which acts in conjunction with the fibres between the two pyramid cartilages (*vide* fig. 8). I must also direct especial attention to a muscle belonging to the adductor group, which has another important function especially related to vocalisation: it is sometimes called the vocal muscle; it runs from the pyramid cartilage to the shield cartilage; it apparently consists of two portions, an external, which acts with the lateral ring-shield muscle and helps to approximate the vocal cords; and another portion situated within the vocal cord itself, which by contracting shortens the vocal cord and probably allows only the free edge to vibrate; moreover, when not contracting, by virtue of the perfect elasticity of muscle the whole thickness of the cord, including this vocal muscle, can be stretched and thrown into vibration (*vide* fig. 8). In the production of chest notes the whole vocal cord is vibrating, the difference in the pitch depending upon the tension produced by the contraction of the tensor (ring-shield) muscle. When, however, the change from the lower to the upper register occurs, as the photographs taken by Dr. French and reproduced in a lecture at the Royal Institution by Sir Felix Semon show, the vocal cords become shorter, thicker, and rounder; and this can be explained by supposing that the inner portion of the vocal muscle contracts at the break from the lower to the upper register (*vide* fig. 11); and that as a result only the free edges of the cords vibrate, causing a change in the quality of the tone. As the scale is ascended the photographs show that the cords become longer and tenser, which we may presume is due to the continued action of the tensor muscle. Another explanation is possible, viz. that in the lower register the two edges of the vocal cords are comparatively thick strings. When the break occurs, owing to the contraction of the inner portion of the vocal muscle, we have a transformation into thin strings, at first short, but as the pitch of the note rises, the thin string formed by the edge of the vocal cord is stretched and made longer by the tensor. It should be mentioned that Aikin and many other good authorities do not hold this view.

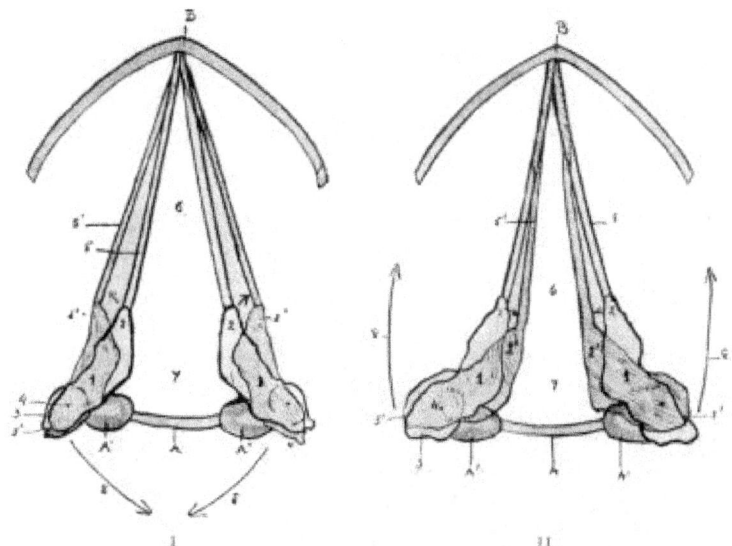

Fig. 7

A-A', Ring Cartilage. B, Shield Cartilage. 1, Pyramid Cartilage. 2, Vocal Process. With 2', Its Position After Contraction of Muscle. 3, Postero-External Base of Pyramid, Giving attachment to Abductor and Adductor Muscles at Rest, With 3', Its New Position After Contraction of the Muscles. 4, Centre of Movement of the Pyramid Cartilage. 5, the Vocal Cords at Rest. 5', their New Position After Contraction of the Abductor and Adductor Muscles, Respectively Seen in I and II. 6, the interligamentous, With 7, the intercartilaginous Chink of the Glottis. 8, the Arrow indicating Respectively in I and II the Action of the Abductor and Adductor in Opening and Closing the Glottis.

FIG. 7.—Diagram after Testut (modified), showing: (i.) the action of the abductor muscle upon the pyramid cartilages in separating the vocal cords; (ii.) the action of the adductor muscles in approximating the vocal cords.

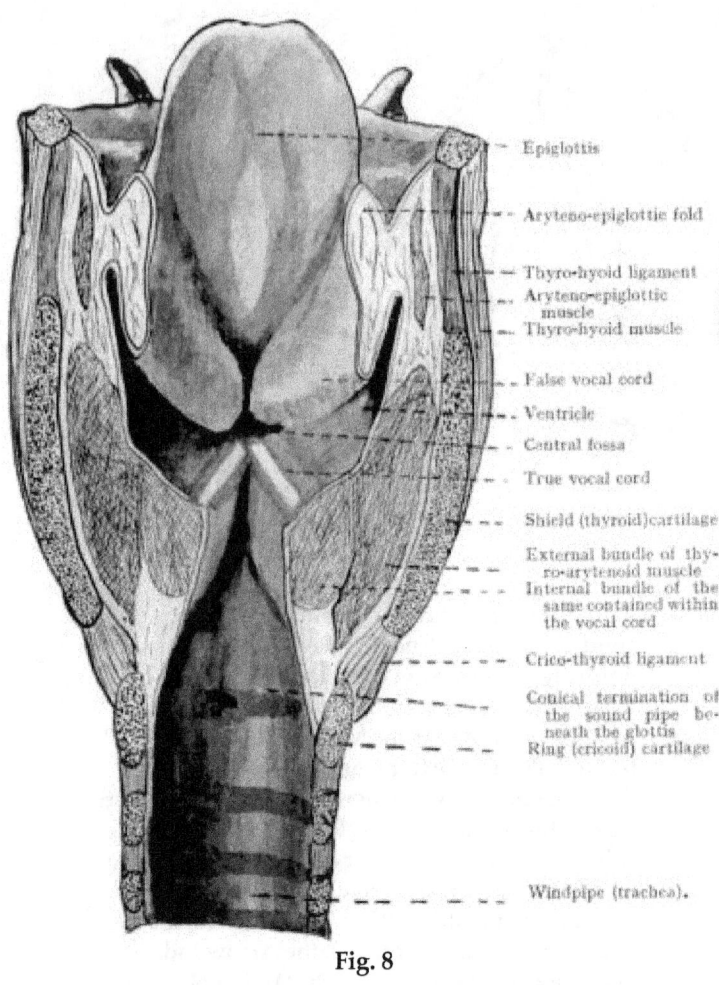

Fig. 8

FIG. 8.—Diagram after Testut (modified) with hinder portion of larynx and windpipe cut away, showing the conical cavity of the sound-pipe below the vocal cords. The ventricle above the vocal cords is seen with the surface sloping upwards towards the mid line.

A diagram showing a vertical section through the middle of the larynx at right angles to the vocal cords shows some important facts in connection with the mechanism of this portion of the vocal instrument (*vide* fig. 8). It will be observed that the sound-pipe just beneath the membranous reed assumes the form of a cone, thus the expired air is driven like a wedge against the closed glottis. Another fact of importance may be observed, that above the vocal cords on either side is a pouch called a ventricle, and the upper surfaces of the vocal cords slope somewhat upwards from without inwards, so that the pressure of the air from above tends to press the edges together. The force of the expiratory blast of air from below overcomes the forces which approximate the edges of the cords and throws them into vibration. With each vibration of the membranous reeds the valve is opened, and as in the case of the siren a little puff of air escapes; thus successive rhythmical undulations of the air are produced, constituting the sound waves. The pitch of the note depends upon the number of waves per second, and the *register* of the voice therefore depends upon two factors: (1) the size of the voice-box, or larynx, and the length of the cords, and (2) the action of the neuro-muscular mechanism whereby the length, approximation, and tension of the vocal cords can be modified when singing from the lowest note to the highest note of the register.

Thus the compass of the—

> Bass voice is D to f 75- 354 vibs. per sec. Tenor " c " c" 133- 562 " " Contralto " e " g" 167- 795 " " Soprano " b " f''' 239-1417 " "

The complete compass of the human voice therefore ranges from about D 75 to f''' 1417 vibrations per second, but the quality of the same notes varies in different individuals.

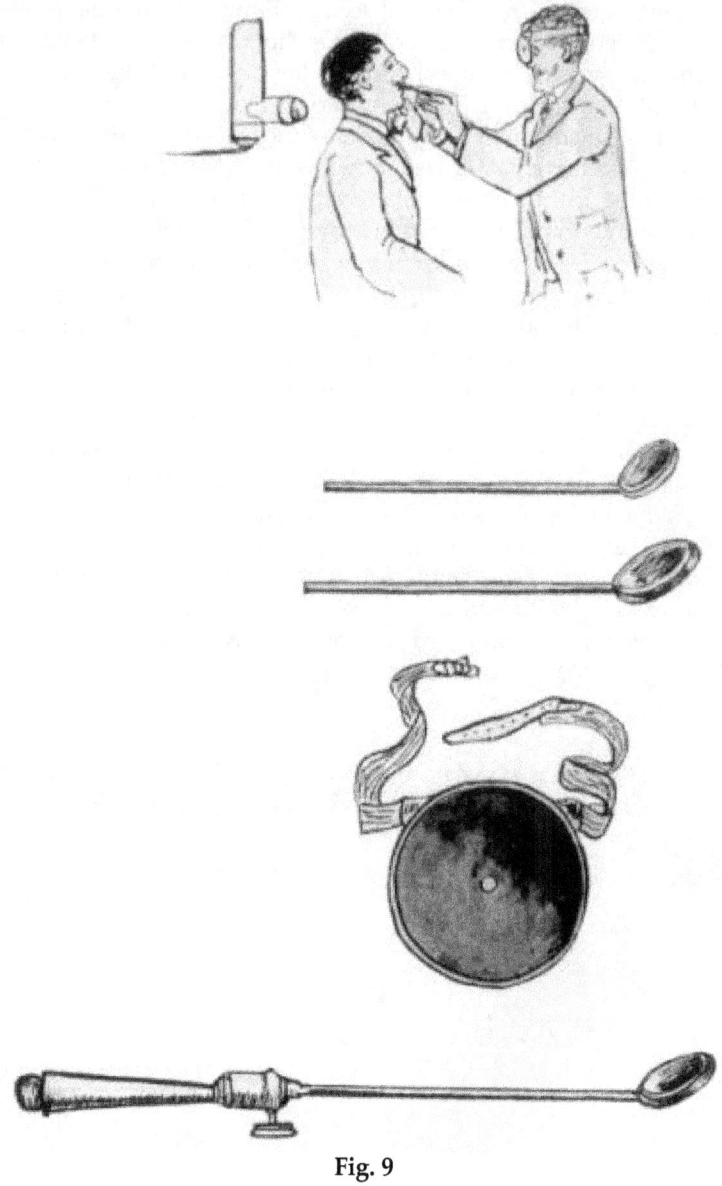

Fig. 9

Fig. 9. — *Description of the laryngoscope and its mode of use.* — The laryngoscope consists of a concave mirror which is fixed on the forehead with a band in such a way that the right eye looks through the hole in the middle. This mirror reflects the light from a lamp placed behind the right side of the patient, who is told to open the mouth and put out the tongue. The observer holds the tongue out gently with a napkin and reflects the light from the mirror on his forehead on to the back of the throat. The small mirror, set at an angle of 450 with the shaft, is of varying size, from half an inch to one inch in diameter, and may be fixed in a handle according to the size required. The mirror is warmed to prevent the moisture of the breath obscuring the image, and it is introduced into the back of the throat in such a manner that the glottis appears reflected in it. The light from the lamp is reflected by the concave mirror on to the small mirror, which, owing to its angle of 450, illuminates the glottis and reflects the image of the glottis with the vocal cords.

The discovery of the laryngoscope by Garcia enabled him by its means to see the vocal cords in action and how the reed portion of the vocal instrument works (*vide* fig. 9 and description). The chink of the glottis or the opening between the vocal cords as seen in the mirror of the laryngoscope varies in size. The vocal cords or ligaments appear dead white and contrast with the surrounding pink mucous membrane covering the remaining structures of the larynx. Fig. 10 shows the appearance of the glottis in respiration and vocalisation. The vocal cords of a man are about seven-twelfths of an inch in length, and those of a boy (before the voice breaks) or of a woman are about five-twelfths of an inch; and there is a corresponding difference in size of the voice-box or larynx. This difference in length of the vocal cords accounts for the difference in the pitch of the speaking voice and the register of the singing voice of the two sexes. We should also expect a constant difference in the length of the cords of a tenor and a bass in the male, and of the contralto and soprano in the female, but such is not the case. It is not possible to determine by laryngoscopic examination what is the natural register of an individual's voice. The vocal cords may be as long in the tenor as in the bass; this shows what an important part the resonator plays in the timbre or quality of the voice. Still, it is generally speaking true, that a small larynx is more often associated with a higher pitch of voice than a large larynx.

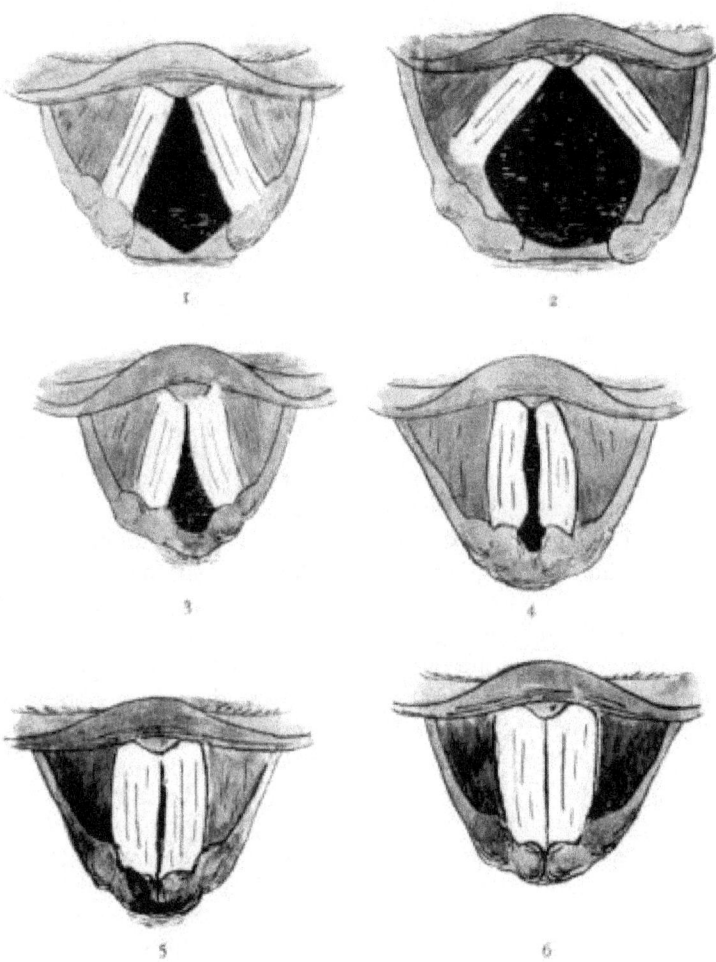

Fig. 10

Fig. 10. — Diagram (modified from Aikin) illustrating the condition of the vocal cords in respiration, whispering, and phonation. (1) Ordinary breathing; the cords are separated and the windpipe can be seen. (2) Deep inspiration; the cords are widely separated and a greater extent of the windpipe is visible. (3) During the whisper the vocal cords are separated, leaving free vent for air through the glottis; consequently there is no vibration and no sound produced by the cords. (4) The soft vocal note, or aspirate, shows that the chink of the glottis is not completely closed, and especially the rima respiratoria (the space between the vocal processes of the pyramidal cartilages.) (5) Strong vocal note, produced in singing notes of the lower register. (6) Strong vocal note, produced in singing notes of the higher register.

Musical notes are comprised between 27 and 4000 vibrations per second. The extent and limit of the voice may be given as between C 65 vibrations per second and f''' 1417 vibrations per second, but this is most exceptional, it is seldom above c''' 1044 per second. The compass of a well-developed singer is about two to two and a half octaves. The normal pitch, usually called the "diapason normal," is that of a tuning-fork giving 433 vibrations per second. Now what does the laryngoscope teach regarding the change occurring in the vocal cords during the singing of the two to two and a half octaves? If the vocal cords are observed by means of the laryngoscope during phonation, no change is *seen*, owing to the rapidity of the vibrations, although a scale of an octave may be sung; in the lower notes, however, the vocal cords are seen not so closely approximated as in the very high notes. This may account for the difficulty experienced in singing high notes piano. Sir Felix Semon in a Friday evening lecture at the Royal Institution showed some remarkable photographs, by Dr. French, of the larynx of two great singers, a contralto and a high soprano, during vocalisation, which exhibit changes in the length of the vocal cords and in the size of the slit between them. Moreover, the photographs show that the vocal cords at the break from the lower to the upper register exhibit characteristic changes.

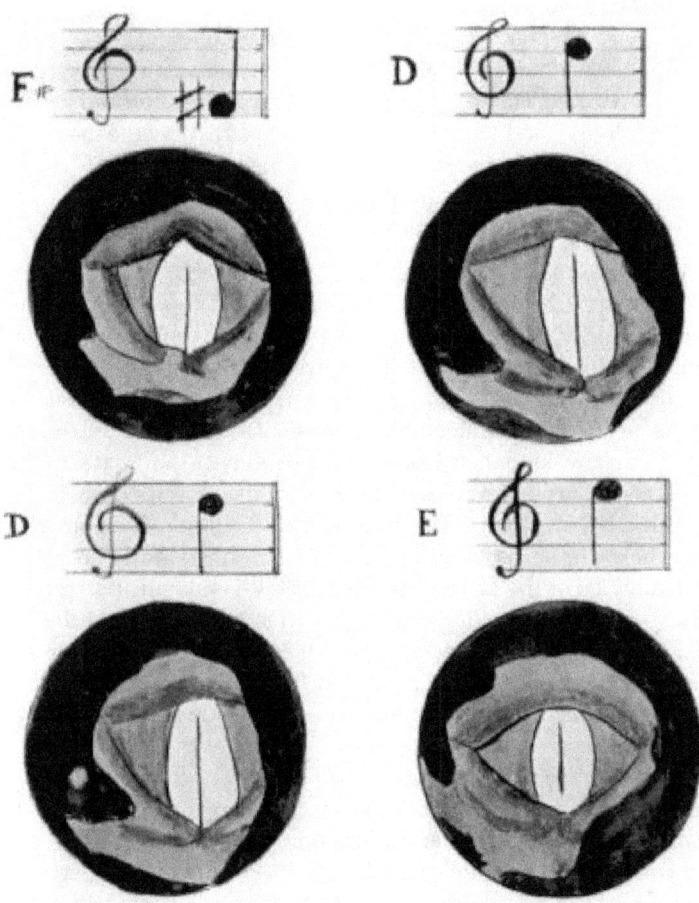

Fig. 11

Fig. 11.—Drawings after Dr. French's photographs in Sir Felix Semon's lecture on the Voice, (1) Appearance of vocal cords of contralto singer when singing F# to D; it will be observed that the cords increase in length with the rise of the pitch, presumably the whole cord is vibrating, including the inner strand of the vocal muscle. At the break from D to E (3 and 4) the cords suddenly become shorter and thicker; presumably the inner portion of the vocal muscle (thyro-arytenoid) is contracting strongly, permitting only the edge of the cord to vibrate. For the next octave the cords are stretched longer and longer; this may be explained by the increasing force of contraction of the tensor muscle stretching the cords and the contained muscle, which is also contracted.

When we desire to produce a particular vocal sound, a mental perception of the sound, which is almost instinctive in a person with a musical ear, awakens by association motor centres in the brain that preside over the innervation currents necessary for the approximation and minute alterations in the tensions of the vocal cords requisite for the production of a particular note. We are not conscious of any kinfsthetic (sense of movement) guiding sensations from the laryngeal muscles, but we are of the muscles of the tongue, lips, and jaw in the production of articulate sounds. It is remarkable that there are hardly any sensory nerve endings in the vocal cords and muscles of the larynx, consequently it is not surprising to find that the ear is the guiding sense for correct modulation of the loudness and pitch of the speaking as well as the singing voice. In reading music, visual symbols produced by one individual awakens in the mind of another mental auditory perceptions of sound varying in pitch, duration, and loudness. Complex neuro-muscular mechanisms preside over these two functions of the vocal instrument. The instrument is under the control of the will as regards the production of the notes in loudness and duration, but not so as regards pitch; for without the untaught instinctive sense of the mental perception of musical sounds correct intonation cannot be obtained by any effort of the will. The untaught ability of correct appreciation of variations in the pitch of notes and the memorising and producing of the same vocally are termed a musical ear. A gift even to a number of people of poor intelligence, it may or may not be associated with the sense of rhythm, which, as we have seen, is dependent upon the mental perception of successive movements associated with a sound. Both correct modulation and rhythm are essential for melody. The sense of hearing is the primary incitation to the voice. This accounts for the fact that children who have learnt to speak, and suffer in early life with ear disease, lose the use of their vocal instrument unless they are trained by lip language and imitation to speak. The remarkable case of Helen Keller, who was born blind and deaf, and yet learned by the tactile motor sensibility of the fingers to feel the vibrations of the vocal organ and translate the perceptions of these vibrations into movements of the lips and tongue necessary for articulation, is one of the most remarkable facts in physiological psychology. Her voice, however, was monotonous, and lacked the modulation in pitch of a musical voice. Music meant little to her but beat and pulsation. She could not sing and she could not play the piano. The fact that Beethoven composed some of his grandest symphonies when stone deaf shows the extraordinary musical faculty he must have preserved to bear in his mind the grand harmonies that he associated with visual symbols. Still, it is impossible that Beethoven, had he been deaf in his early childhood, could ever have developed into the great musical genius that he became.

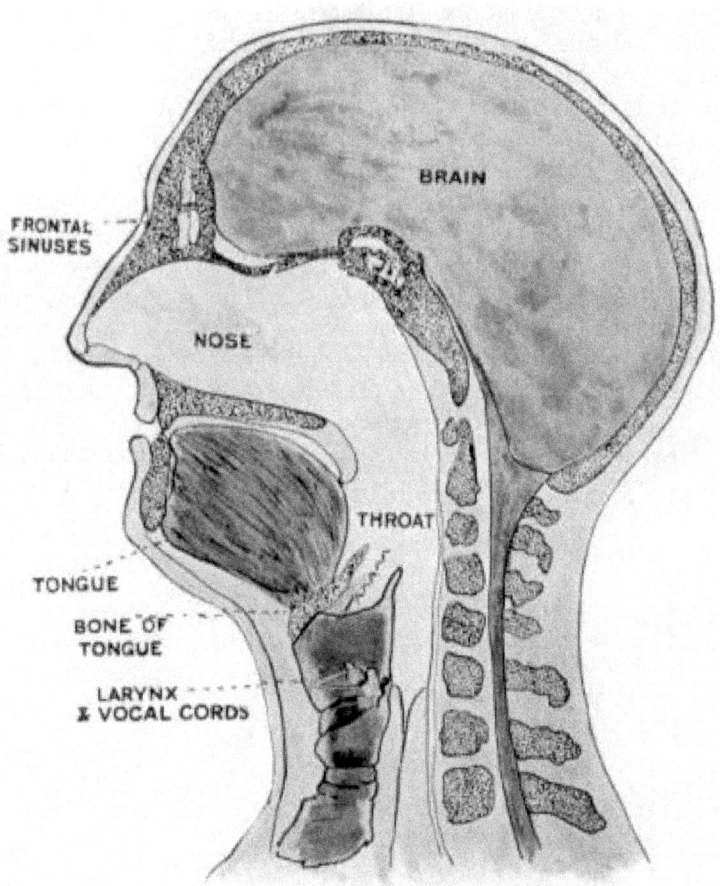

Fig. 12

Fig. 12.—Diagram showing the position of the larynx in respect to the resonator and tongue. The position of the vocal cords is shown, but really they would not be seen unless one half of the shield cartilage were cut away so as to show the interior of the voice-box. Sound vibrations are represented issuing from the larynx, and here they become modified by the resonator; the throat portion of the resonator is shown continuous with the nasal passages; the mouth portion of the resonator is not in action, owing to the closure of the jaw and lips. The white spaces in the bones of the skull are air sinuses. In such a condition of the resonator, as in humming a tune, the sound waves must issue by the nasal passages, and therefore they acquire a nasal character.

III. THE RESONATOR AND ARTICULATOR

The Resonator.—The resonator is an irregular-shaped tube with a bend in the middle; the vertical portion is formed by the larynx and pharynx, the horizontal by the mouth. The length of the resonator, from the vocal cords to the lips, is about 6.5 to 7 inches (*vide* fig. 12). The walls of the vertical portion are formed by the vertebral column and the muscles of the pharynx behind, the cartilages of the larynx and the muscles of the pharynx at the sides, and the thyroid cartilage, the epiglottis, and the root of the tongue in front; these structures form the walls of the throat and are all covered with a mucous membrane. This portion of the resonator passage can be enlarged to a slight degree by traction upon the larynx below (sterno-thyroid muscle), by looseness of the pharyngeal muscles, and still more by the forward placement of the tongue; the converse is true as regards diminution in size. The horizontal portion of the resonator tube (the mouth) has for its roof the soft palate and the hard palate, the tongue for its floor, and cheeks, lips, jaw, and teeth for its walls. The interior dimensions of this portion of the resonator can be greatly modified by movements of the jaw, the soft palate, and the tongue, while the shape and form of its orifice is modified by the lips.

There are accessory resonator cavities, and the most important of these is the nose; its cavity is entirely enclosed in bone and cartilage, consequently it is immovable; this cavity may or may not be closed to the sonorous waves by the elevation of the soft palate. When the mouth is closed, as in the production of the consonant m, e.g. in singing *me*, a nasal quality is imparted to the voice, and if a mirror be placed under the nostrils it will be seen by the vapour on it that the sound waves have issued from the nose; consequently the nasal portion of the resonator has imparted its characteristic quality to the sound. The air sinuses in the upper jaws, frontal bones, and sphenoid bones act as accessory resonators; likewise the bronchi, windpipe, and lungs; but all these are of lesser importance compared with the principal resonating chamber of the mouth and throat. If the mouth be closed and a tune be hummed the whole of the resonating chambers are in action, and the sound being emitted from the nose the nasal quality is especially marked. But no sound waves are produced unless the air finds an exit; thus a tune cannot be hummed if both mouth and nostrils are closed.

From the description that I have given above, it will be observed that the mouth, controlled by the movements of the jaw, tongue, and lips, is best adapted for the purpose of articulate speech; and that the throat, which is less actively movable and contains the vocal cords, must have greater influence on the sound vibrations without participating in the articulation of words. While the vocal cords serve the purpose of the reed, the resonator forms the body of the vocal instrument. Every sound passes through it; every vowel and consonant in the production of syllables and words must be formed by it, and the whole character and individual qualities of the speaking as well as the singing voice depend in great part upon the manner in which it is used.

The acoustic effect is due to the resonances generated by hollow spaces of the resonator, and Dr. Aikin, in his work on "The Voice," points out that we can study the resonances yielded by these hollow spaces by whispering the vocal sounds; but it is necessary to put the resonator under favourable conditions for the most efficient production. When a vowel sound is whispered the glottis is open (*vide* fig. 10) and the vocal cords are not thrown into vibration; yet each vowel sound is associated with a distinct musical note, and we can produce a whole octave by alteration of the resonator in whispering the vowel sounds. In order to do this efficiently it is necessary to use the bellows and the resonator to the best advantage; therefore, after taking a deep inspiration in the manner previously described, the air is expelled through the open glottis into the resonating cavity, which (as fig. 13 shows) is placed under different conditions according to the particular vowel sound whispered. In all cases the mouth is opened, keeping the front teeth about one inch apart; the tongue should be in contact with the lower dental arch and lie as flat on the floor of the mouth as the production of the particular vowel sound will permit. When this is done, and a vowel sound whispered, a distinctly resonant note can be heard. Helmholtz and a number of distinguished German physicists and physiologists have analysed the vowel sounds in the whispering voice and obtained very different results. If their experiments show nothing else, they certainly indicate that there are no universally fixed resonances for any particular vowel sound. Some of the discrepancies may (as Aikin points out) be due to the conditions of the experiment not being conducted under the same conditions. Aikin, indeed, asserts that if the directions given above be fulfilled, there will be variations between full-grown men and women of one or two tones, and between different men and different women of one or two semi-tones, and not much more. As he truly affirms, if the tube is six inches long a variation of three-quarters of an inch could only make a difference of a whole tone in the resonance, and he implies that the different results obtained by these different experimenters were due to the faulty use of the resonator.

In ordinary conversation much faulty pronunciation is overlooked so long as the words themselves are intelligible, but in singing and public speaking every misuse of the resonator is magnified and does not pass unnoticed. Increased loudness of the voice will not improve its carrying power if the resonator is improperly used; it will often lead to a rise of pitch and the production of a harsh, shrill tone associated with a sense of strain and effort. Aikin claims that by studying the whispering voice we can find for every vowel sound that position of the resonator which gives us the maximum of resonance. By percussing2 the resonator in the position for the production of the various vowel sounds you will observe a distinct difference in the pitch of the note produced. I will first produce the vowel sound *oo* and proceed with the vowel sounds to *i*; you will observe that the pitch rises an octave; that this is due to the changes in the form of the resonator is shown when I percuss the resonator in the position of the different vowel sounds. You will observe that I start the scale of C with *oo* on f and proceed through a series of vowel sounds as in whispering *who, owe, or, on, ah*. I rise a fifth from f to c, and the diagram shows the change in the form of the resonator cavity to be mainly due to the position of the dorsum of the tongue. Proceeding from *ah* to the middle tone of the speaking register, we ascend the scale to *i* as in *me*, and the dorsum of the tongue now reaches the roof of the mouth; but the tongue not only rises, it comes forward, and the front segment of the resonator is made a little smaller at every step of the scale while the back segment becomes a little larger. I consider this diagram of Aikin to be more representative of the changes in the resonator than the description of Helmholtz, who stated that the form of the resonator during the production of the vowel sound *u* and *o* is that of a globular flask with a short neck; during the production of *a* that of a funnel with the wide extremity directed forward; of *e* and *i* that of a globular flask with a long narrow neck.

> [Footnote 2: This was done by the lecturer placing his left forefinger on the outside of the right cheek, then striking it with the tip of the middle finger of the right hand, just in the same way as he would percuss the chest. — F.W.M.]

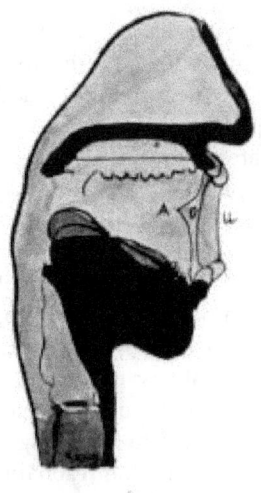 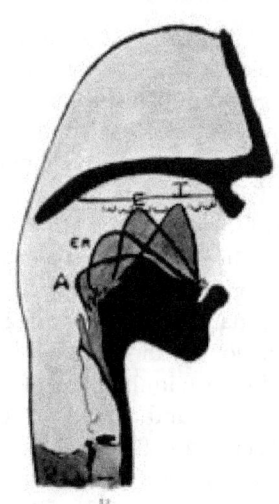

Fig. 13

FIG. 13. — Diagram after Aikin.
1. To show position of tongue and lips in the production of the vowel sounds *a, o, oo*.
2. To show successive positions of the tongue in the production of the vowel sounds *a, ei, e, i*.

I have already said that Helmholtz showed that each vowel sound has its particular overtones, and the quality or "timbre" of the voice depends upon the proportional strength of these overtones. Helmholtz was able by means of resonators to find out what were the overtones for each vowel sound when a particular note was sung. The flame manometer of Kvnig (*vide* fig. 14) shows that if the same note be sung with different vowels the serrated flame image in the mirror is different for each vowel, and if a more complicated form of this instrument be used (such as I show you in a picture on the screen) the overtones of the vowel sounds can be analysed. You will observe that this instrument consists of a number of resonators placed in front of a series of membranes which cover capsules, each capsule being connected with a jet of gas.

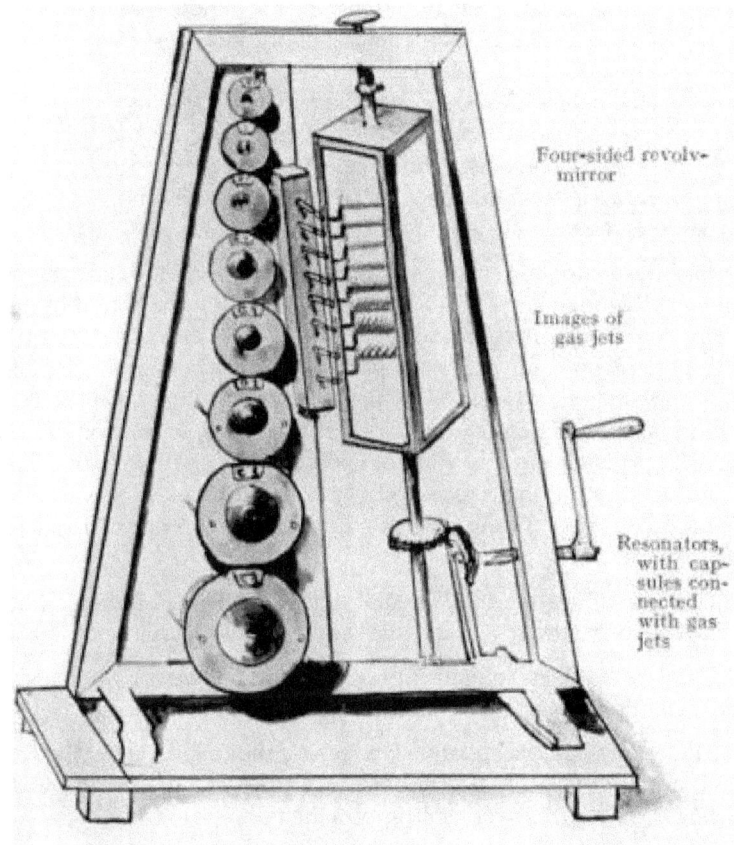

Fig. 14

FIG. 14.—Kvnig's flame manometer. The fundamental note C is sung on a vowel sound in front of the instrument; the lowest resonator is proper to that note and the air in it is thrown into corresponding periodic rhythmical vibrations, which are communicated through an intervening membrane to the gas in the capsule at the back of the resonator; but the gas is connected with the lighted jet, the flame of which is reflected in the mirror, the result being that the flame vibrates. When the mirror is made to revolve by turning the handle the reflected image shows a number of teeth corresponding to the number of vibrations produced by the note which was sung. The remaining resonators of the harmonic series with their capsules and gas-jets respond in the same manner to the overtones proper to each vowel sound when the fundamental note is sung.

Each resonator corresponds from below upwards to the harmonics of the fundamental note c. In order to know if the sound of the voice contains harmonics and what they are, it is necessary to sing the fundamental note c on some particular vowel sound; the resonators corresponding to the particular harmonics of the vowel sound are thus set in action, and a glance at the revolving mirror shows which particular gas jets vibrate. Experiments conducted with this instrument show that the vowel U=*oo* is composed of the fundamental note very strong and the third harmonic (viz. g) is fairly pronounced.

O (*on*) contains the fundamental note, the second harmonic (the octave c') very strong, and the third and fourth harmonics but weak.

The vowel A (*ah*) contains besides the fundamental note, the second harmonic, weak; the third, strong; and the fourth, weak.

The vowel E (*a*) has relatively a feeble fundamental note, the octave above, the second harmonic, is weak, and the third weak; whereas the fourth is very strong, and the fifth weak.

The vowel I (*ee*) has very high harmonics, especially the fifth, which is strongly marked.

We see from these facts that there is a correspondence between the existence of the higher harmonics and the diminished length of the resonator. They are not the same in all individuals; for they depend also upon the *timbre* of the voice of the person pronouncing them, or the special character of the language used, as well as upon the pitch of the fundamental notes employed.

Helmholtz inferred that if the particular quality of the vowel sounds is due to the reinforcement of the fundamental tone by particular overtones, he ought to be able to produce synthetically these vowel sounds by combining the series of overtones with the fundamental note. This he actually accomplished by the use of stopped organ pipes which gave sensibly simple notes.

Having thus shown that the fundamental note is dependent upon the tension of the vocal cords—the reed portion of the instrument—and the quality, timbre, or "klang" upon the resonator, I will pass on to the formation of syllables and words of articulate speech by the combination of vowel sounds and consonants.

"The articulate sounds called consonants are sounds produced by the vibrations of certain easily movable portions of the mouth and throat; and they have a different sound according as they are accompanied by voice or not" (Hermann).

The emission of sounds from the resonator may be modified by interruption or constriction in three situations, at each of which added vibrations may occur, (1) At the lips, the constriction being formed by the two lips, or by the upper or lower lip with the lower or upper dental arch. (2) Between the tongue and the palate, the constriction being caused by the opposition of the tip of the tongue to the anterior portion of the hard palate or the posterior surface of the dental arch. (3) At the fauces, the constriction being due to approximation of the root of the tongue and the soft palate. Consonants can only be produced in conjunction with a vowel sound, consequently the air is thrown into sonorous waves of a complex character, in part dependent upon the shape of the resonator for the production of the vowel, in part dependent upon the vibrations at each of these situations mentioned above. Consonants may accordingly be classified as they are formed at the three places of interruption—lips, teeth, and fauces respectively: (1) labial; (2) dental; (3) guttural.

The sounds formed at each of the places of interruption are divided into—

1. *Explosives.*—At one of the situations mentioned the resonator is suddenly opened or closed during the expulsion of air—(*a*) without the aid of voice, p, t, k; (*b*) with the aid of voice, b, d, g. When one of these consonants begins a syllable, opening of the resonator is necessary, e.g. pa; when it ends a syllable, closure is necessary, e.g. ap. No sharp distinction is possible between p and b, t and d, and k and g if they are whispered.

2. *Aspirates.*—The resonator is constricted at one of the points mentioned so that the current of air either expired or inspired rushes through a small slit. Here again we may form two classes: (*a*) without the aid of the voice, f, s (sharp), ch, guttural; (*b*) with the aid of voice, v, z, y. The consonants s and l are formed when the passage in front is closed by elevation of the tongue against the upper dental arch so that the air can only escape at the sides between the molar teeth: sh is formed by the expulsion of the current of air through two narrow slits, viz. (1) between the front of the tongue and the hard palate, the other between the nearly closed teeth. If a space be left between the tip of the tongue and the upper teeth two consonant sounds can be produced, one without the aid of the voice—th (hard) as in that; the other with the aid of voice—th (soft) as in thunder. Ch is a guttural produced near the front of the mouth, e.g. in Christ, or near the back as in Bach.

3. *Resonants.*—In the production of the consonant m, and sometimes n, the nasal resonator comes into play because the soft palate is not raised at all and the sound waves produced in the larynx find a free passage through the nose, while the mouth portion of the resonator is completely closed by the lips. The sounds thus produced are very telling in the singing voice.

4. *Vibratory Sounds.*—There are three situations in which the consonant r may be formed, but in the English language it is produced by the vibration of the tip of the tongue in the constricted portion of the cavity of the mouth, formed by the tongue and the teeth.

The consonants have been grouped by Hermann as follows:—

	Labials.	**Dentals.**	**Gutturals.**
1. Explosives—			
a. Without the voice	P	T	K
b. With the voice	B	D	G
2. Aspirates—			
a. Without the voice	F	S (hard), L, Sh, Th (hard)	Ch
b. With the voice	V	Z, L, Th, Zh (soft)	Y in yes
3. Resonants	M	N	N (nasal)
4. Vibratory sounds	Labial R	Lingual R	Guttural R

H is the sound produced in the larynx by the quick rushing of the air through the widely opened glottis.

Hermann's classification which I have given is especially valuable as regards the speaking voice, but Aikin classifies the consonants from the singing point of view, according to the more or less complete closure of the resonator.

CLASSIFICATION OF CONSONANTS (AIKIN)

Jaw fully open	H, L, K, G
Jaw less open	T, D, N, R
Jaw nearly closed, lips closed	P, B, M
Jaw nearly closed, upper lip on lower teeth	F, V
Jaw quite closed	S, Z, J, N, Ch, Sh

Aikin, moreover, points out that the English language is so full of closures that it is difficult to keep the resonator open, and that accounts for one of the principal difficulties in singing it.

"The converse of this may be said of Italian, in which most words end in pure vowels which keep the resonator open. In fact, it is this circumstance which has made the Italian language the basis of every point of voice culture and the producer of so many wonderful singers." As an example compare the English word 'voice,' which begins with closure and ends with closure, and the Italian 'voce,' pronounced *vochi*, with its two open vowel sounds. The vowel sound ah on the note c is the middle tone of the speaking register, and as we know, can be used all day long without fatigue; therefore in training the voice the endeavour should be made to develop the register above and below this middle tone. In speaking there is always a tendency under emotional excitement, especially if associated with anger, to raise the pitch of the voice, whereas the tender emotions lead rather to a lowering of the pitch. Interrogation generally leads to a rise of the pitch; thus, as Helmholtz pointed out, in the following sentence there is a decided fall in the pitch — "I have been for a walk"; whereas in "Have you been for a walk?" there is a decided rise of pitch. If you utter the sentence "Who are you?" there is a very definite rise of pitch on 'are.'

PATHOLOGICAL DEGENERATIVE CHANGES PRODUCING SPEECH DEFECTS AND WHAT THEY TEACH

As I have before remarked, children utter vowel sounds before consonants, and I used this as an argument that phonation preceded articulation; but there is another reason for supposing that articulate sounds are of later development phylogenetically, as well as ontogenetically. Not only are they more dependent for their proper production on intelligence, but in those disorders of speech which occur as a result of degenerative processes of the central nervous system the difficulty of articulate speech precedes that of phonation. Take, for example, bulbar paralysis, a form of progressive muscular atrophy, a disease due to a progressive decay and destruction of the motor nerve cells presiding over the movements of the tongue, lips, and larynx, hence often called glosso-labial-laryngeal palsy. In this disease the lips, tongue, throat, and often the larynx are paralysed on both sides. "The symptoms are, so to speak, grouped about the tongue as a centre, and it is in this organ that the earliest symptoms are usually manifested." (Gowers). Imperfect articulation of those sounds in which the tongue is chiefly concerned, viz. the lingual consonants l, r, n, and t, causing indistinctness of speech, is the first symptom; the lips then become affected and there is difficulty in the pronunciation of sounds in which the lips are concerned, viz. u, o, p, b, and m. Eventually articulate speech becomes impossible, and the only expression remaining to the patient is laryngeal phonation, slightly modulated and broken into the rhythm of formless syllables.

The laryngeal palsy *rarely* becomes complete. The nervous structures in the *physiological mechanism* of speech and phonation are affected in this disease; but there are degenerative diseases of the brain in which the *psychical mechanism* of speech is affected, e.g. General Paralysis of the Insane, in which the affection of speech and hand-writing is quite characteristic. There is at first a hesitancy which may only be perceptible to practised ears, but in which there is no real fault of articulation once it is started; sometimes preparatory to and during the utterance there is a tremulous motion about the muscles of the mouth. The hesitation increases, and instead of a steady flow of modulated, articulate sounds, speech is broken up into a succession of irregular, jerky, syllabic fragments, without modulation, and often accompanied by a tremulous vibration of the voice. Syllables are unconsciously dropped out, blurred, or run into one another, or imperfectly uttered; especially is difficulty found with consonants, particularly explosive sounds, b, p, m; again, linguals and dentals are difficult to utter. Similar defects occur in written as in vocal speech; the syllables and even the letters are disjointed; there is a fine tremor in the writing, and inco-ordination in the movements of the pen. Silent thoughts leave out syllables and words in the framing of sentences; consequently they are not expressed by the hand. The ideation of a written or spoken word is based upon the association of the component syllables, and the difficulty arises primarily from the progressive impairment of this function of association upon which spoken and written language so largely depends. Examination of the brain in this disease explains the cause of the speech trouble and the progressive dementia (loss of mind) and paralysis with which it is associated. There is a wasting of the cerebral hemispheres, especially of the frontal lobes, a portion of the brain which, later on, we shall see is intimately associated with the function of articulate speech.

THE CEREBRAL MECHANISM OF SPEECH AND SONG

Neither vocalisation nor articulation are essentially human. Many of the lower animals, e.g. parrots, possess the power of articulate speech, and birds can be taught to pipe tunes. The essential difference between the articulate speech of the parrot and the human being is that the parrot merely imitates sounds, it does not employ these articulate sounds to express judgments; likewise there are imbecile human beings who, parrot-like, repeat phrases which are meaningless. Articulate speech, even when employed by a primitive savage, always expresses a judgment. Even in the simple psychic process of recalling the name aroused by the sight of a common object in daily use, and in affixing the verbal sign to that object, a judgment is expressed. But that judgment is based upon innumerable experiences primarily acquired through our special senses, whereby we have obtained a knowledge of the properties and uses of the object. This statement implies that the whole brain is consciously and unconsciously in action. There is, however, a concentration of psychic action in those portions of the brain which are essential for articulate speech; consequently the word, as it is mentally heard, mentally seen, and mentally felt (by the movements of the jaw, tongue, lips, and soft palate), occupies the field of clear consciousness; but the concept is also the nucleus of an immense constellation of subconscious psychic processes with which it has been associated by experiences in the past. In language, articulate sounds are generally employed as objective signs attached to objects with which they have no natural tie.

In considering the relation of the Brain to the Voice we have not only a physiological but a psychological problem to deal with. Since language is essentially a human attribute, we can only study the relation of the Brain to Speech by observations on human beings who during life have suffered from various speech defects, and then correlate these defects with the anatomical changes found in the brain after death.

Between the vocal instrument of the primitive savage and that of the most cultured singer or orator there is little or no discoverable difference; neither by careful naked-eye inspection of the brain, nor aided by the highest powers of the microscope, should we be able to discover any sufficient structural difference to account for the great difference in the powers of performance of the vocal instrument of the one as compared with that of the other; nor is there any sufficient difference in size or minute structure of the brain to account for the vast store of intellectual experiences and knowledge of the one as compared with the other. The cultured being descended from cultured beings inherits tendencies whereby particular modes of motion or vibration which have been experienced by ancestors are more readily aroused in the central nervous system; when similar stimuli producing similar modes of motion affect the sense organs. But suppose there were an island inhabited only by deaf mutes, upon which a ship was wrecked, and the sole survivors of the wreck were infants who had never used the voice except for crying, would these infants acquire articulate speech and musical vocalisation? I should answer, No. They would only be able to imitate the deaf mutes in their gesture language and possibly the musical sounds of birds; for the language a child learns is that which it hears; they might however develop a simple natural language to express their emotions by vocal sounds. The child of English-speaking parents would not be able spontaneously to utter English words if born in a foreign country and left soon after birth amongst people who could not speak a word of English, although it would possess a potential facility to speak the language of its ancestors and race.

It is necessary, however, before proceeding further, to say a few words explanatory of the brain and its structure, and the reader is referred to figs. 15, 16, 17. The brain consists of (1) the great brain or cerebrum, (2) the small brain or cerebellum, and (3) the stem of the brain, which is continuous with the spinal cord. The cerebro-spinal axis consists of grey matter and white matter. The grey matter covers the surface of the cerebrum and cerebellum, the white matter being internal. The stem of the brain, the medulla oblongata, and the spinal cord, consists externally of white matter, the grey matter being internal. The grey matter consists for the most part of nerve cells (ganglion cells), and the white matter consists of nerve fibres; it is white on account of the phosphoretted fatty sheath—myelin—

that covers the essential axial conducting portion of the nerve fibres. If, however, the nervous system be examined microscopically by suitable staining methods, it will be found that the grey and white matters are inseparably connected, for the axial fibres of the nerves in the white matter are really prolongations of the ganglion cells of the grey matter; in fact the nervous system consists of countless myriads of nervous units or neurones; and although there are structural differences in the nervous units or neurones, they are all constructed on the same general architectural plan (*vide* fig. 15). They may be divided into groups, systems, and communities; but there are structural differences of the separate systems, groups, and communities which may be correlated with differences of function. The systems may be divided into: (1) afferent sensory, including the special senses and general sensibility; (2) motor efferent; (3) association.

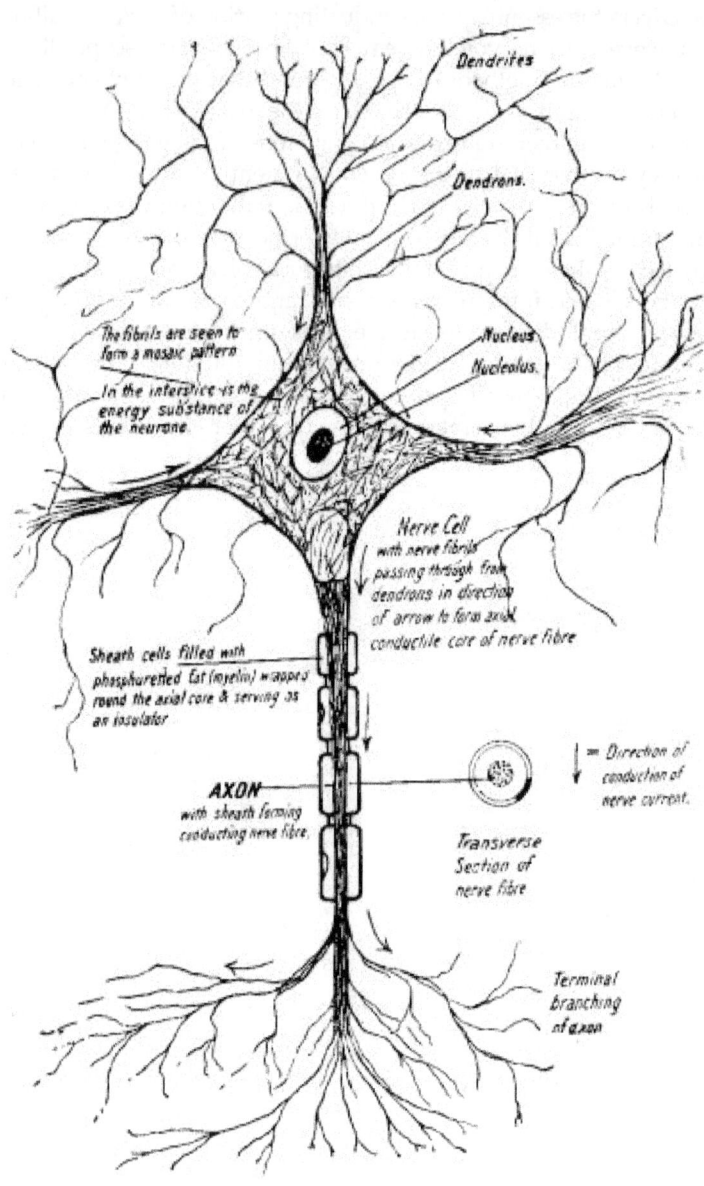

Fig. 15

FIG. 15. — Diagrammatic representation of a motor neurone magnified 300 diameters. Whereas the nerve cell and its branching processes (the dendrons) form but a minute speck of protoplasm, the nerve fibre which arises from it, although microscopic in diameter, extends a very long distance; in some cases it is a yard long; consequently only a minute fraction of the nerve fibre is represented in the diagram.

The great brain or cerebrum consists of two halves equal in weight, termed hemispheres, right and left; and the grey matter covering their surface is thrown into folds with fissures between, thus increasing enormously the superficial area of the grey matter and of the neurones of which it consists without increasing the size of the head. The pattern of the folds or convolutions shows a general similarity in all human beings, certain fissures being always present; and around these fissures which are constantly present are situated fibre systems and communities of neurones having particular functions (*vide* fig. 16.) Thus there is a significance in the convolutional pattern of the brain. But just as there are no two faces alike, so there are no two brains alike in their pattern; and just as it is rare to find the two halves of the face quite symmetrical, so the two halves of the brain are seldom exactly alike in their pattern. Although each hemisphere is especially related to the opposite half of the body, the two are unified in function by a great bridge of nerve fibres, called the corpus callosum, which unites them. The cortical centres or structures with specialised functions localised in particular regions of one hemisphere are associated by fibres passing to the same region in the opposite hemisphere by this bridge.

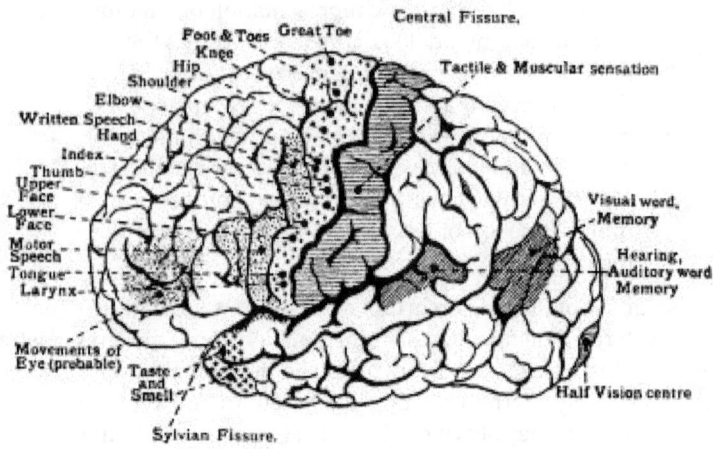

Fig. 16

FIG. 16.—Diagram of the left hemisphere of the brain showing localised centres, of which the functions are known. It will be observed that the centres for the special senses, tactile, muscular, hearing, and vision, are all situated behind the central fissure. The tactile-motor kinfsthetic sense occupies the whole of the post-central convolution; the centre for hearing (and in the left hemisphere memory of words) is shown at the end of the first temporal convolution, but the portion shaded by no means indicates the whole of the grey cortex which possesses this function; a large portion of this centre cannot be seen because it lies within the fissure forming the upper surface of the temporal lobe. Behind this is the angular gyrus which is connected with visual word memory. The half-vision centre, and by this is meant the portion of brain which receives impressions from each half of the field of vision, is situated for the most part on the inner (unseen) surface of the occipital lobe. In front of the central fissure is situated the motor area, or that region destruction of which causes paralysis of the muscles moving the structures of the oppo-

site half of the body. If the situations indicated by
black dots be excited by an interrupted electric
current, movements of the limbs, trunk, and face
occur in the precise order shown, from the great
toe to the larynx. In front of this precentral convolution are the three frontal convolutions, and it
would seem that the functions of these convolutions are higher movements and attention in fixa-
tion of the eyes; moreover, in the lowest frontal region, indicated by fine dots, we have Broca's convolution, which is associated with motor speech;
above at the base of the second middle frontal
convolution is the portion of cortex in which is localised the function of writing. Taste and smell
functions reside in brain cortex only a small portion of which can be seen, viz. that at the tip of the
temporal lobe.

Muscles and groups of muscles on the two sides of the body which invariably act together may thus be innervated from either hemisphere, e.g. the muscles of the larynx, the trunk, and upper part of the face.

Gall, the founder of the doctrine of Phrenology, wrecked his fame as a scientist by associating mental faculties with conditions of the skull instead of conditions of the brain beneath; nevertheless, he deserves the highest credit for his discoveries and deductions, for he was the first to point out that that part of the brain with which psychic processes are connected must be the cerebral hemispheres. He said, if we compare man with animals we find that the sensory functions of animals are much finer and more highly developed than in man; in man, on the other hand, we find intelligence much more highly developed than in animals. Upon comparing the corresponding anatomical conditions, we see, he said, that in animals the deeper situated parts of the brain are relatively more developed and the hemispheres less developed than in man; in man, the hemispheres so surpass in development those of animals that we can find no analogy. Gall therefore argued that we must consider the cerebral hemispheres to be the seat of the higher functions of the mind. We must moreover acknowledge that the following deductions of Gall are quite sound: "The convolutions ought to be recognised as the parts where the instincts, feelings, thoughts, talents, the affective qualities in general, and the moral and intellectual forces are exercised." The Paris Academy of Science appointed a commission of inquiry, May, 1808, which declared the doctrine of Gall to be erroneous. Gall moreover surmised that the faculty of language lay in the frontal lobes, and Bouillaud supported Gall's proposition by citing cases in which speech had been affected during life, and in which after death the frontal lobes were found to be damaged by disease. A great controversy ensued in France; popular imagination was stirred up especially in the republic by the doctrine of Gall, which was an attempt to materialise and localise psychic processes. Unfortunately Gall's imagination, encouraged by a widespread wave of popular sympathy, overstepped his judgment and launched him into speculative hypotheses unsupported by facts. His doctrine of Phrenology was shown to be absolutely illogical; consequently it was forgotten that he was the pioneer of cerebral localisation.

SPEECH AND RIGHT-HANDEDNESS

The next step in Cerebral Localisation was made by a French physician, Marc Dax, who first observed that disease of the left half of the cerebrum producing paralysis of the right half of the body (right hemiplegia) was associated with loss of articulate speech. This observation led to the establishment of a most important fact in connection with speech, viz. that right-handed people use their left cerebral hemisphere as the executive portion of the brain in speech. Subsequently it was shown that when left-handed people were paralysed on the left side by disease of the right hemisphere, they lost their powers of speech. But the great majority of people are born right-handed, consequently the right hand being especially the instrument of the mind in the majority of people, the left hemisphere is the leading hemisphere; and since probably specialisation of function of the right hand (dexterity) has been so closely associated with that other instrument of the mind, the vocal instrument of articulate speech, the two have now become inseparable; for are not graphic signs and verbal signs intimately interwoven in the development of language and human intelligence?

What has determined the predominance of the left hemisphere in speech? I can find no adequate anatomical explanation. There is no difference in weight of the two hemispheres in normal brains. Moreover, I am unable to subscribe to the opinion that there is any evidence to show that the left hemisphere receives a larger supply of blood than the right. Another theory advanced to explain localisation of speech and right-handedness in the left hemisphere is that the heavier organs, lung and liver, being on the right side have determined a mechanical advantage which has led to right-handedness in the great majority of people. This theory has, however, been disposed of by the fact that cases in which there has been a complete transposition of the viscera have not been left-handed in a larger proportion of cases. The great majority of people, modern and ancient, civilised and uncivilised, use the right hand by preference. Even graphic representations on the sun-baked clay records of Assyria, and the drawings on rocks, tusks, and horns of animals of the flint-weapon men of prehistoric times show that man was then right-handed. There is a difference of opinion whether anthropoid apes use the right hand in preference to the left. Professor Cunningham, who made a special study of this subject, asserts that they use either hand indifferently; so also does the infant at first, and the idiot in a considerable number of cases. Then why should man, even primitive, have chosen the right hand as the instrument of the mind? Seeing that there is no apparent anatomical reason, we may ask ourselves the question: Is it the result of an acquired useful habit to which anatomical conditions may subsequently have contributed as a co-efficient? Primitive man depended largely upon gesture language, and the placing of the hand over the heart is universally understood to signify love and fidelity. Uneducated deaf mutes, whose only means of communicating with their fellow-men is by gestures, not only use this sign, but imply hatred also by holding the hand over the heart accompanied by the sign of negation. Moreover, pointing to the heart accompanied by a cry of pain or joy would indicate respectively death of an enemy or friend. Again, primitive man protected himself from the weapons of his enemies by holding the shield in his left hand, thus covering the heart and leaving the right hand free to wield his spear. The question whether it would have been to his advantage to use either hand indifferently for spear and shield has been, to my mind, solved by the fact that in the long procession of ages evolution has determined right-handed specialisation as being more advantageous to the progress of mankind than ambidexterity. Right-handedness is an inherited character in the same sense as the potential power of speech.

LOCALISATION OF SPEECH CENTRES IN THE BRAIN

In 1863 Broca showed the importance in all right-handed people (that is in about ninety-five per cent of all human beings) of the third *left* frontal convolution for speech (*vide* figs. 16 and 17); when this is destroyed by disease, although the patient can understand what is said and can understand written and printed language, the power of articulate speech is lost. *Motor Aphasia*. This portion of the brain is concerned with the revival of the motor images, and has been termed by Dr. Bastian "the glosso-kinfsthetic centre," or the cortical grey matter, in which the images of the sense of movement of the lips and tongue are formed (*vide* fig. 17). A destruction of a similar portion of the cortex in a right-handed person produces no loss of speech; but if the person is left-handed there is aphasia, because he, being left-handed, uses the third *right* inferior frontal convolution for speech. These facts have for long been accepted by most neurologists, but recently doubts have been cast upon this fundamental principle of cerebral localisation by a most distinguished French neurologist, M. Marie; he has pointed out that a destructive lesion of the cortex may be accompanied by subcortical damage, which interrupts fibres coming from other parts of the brain connected with speech.

In the study of speech defects it is useful to employ a diagram; a certain part of the brain corresponds to the *Speech Zone* there indicated, and lesions injuring any part of this area in the left hemisphere cause speech defects (*vide* fig. 17). All neurologists, M. Marie included, admit this, and the whole question therefore is: Is a destruction of certain limited regions of the superficial grey matter the cause of different forms of speech defects, or are they not due more to the destruction of subcortical systems of fibres, which lie beneath this cortical speech zone?

There is a certain portion of the speech zone which is assumed to be connected with the revival of written or printed language, and is called the *visual word-centre*. There is another region connected with the memory of spoken words—the *auditory word-centre*; you will observe that it is situated in the posterior third of the first temporal convolution, but this does not comprise nearly the whole of it, for there is an extensive surface of grey matter lying unseen within the fissure, called the transverse convolutions, or gyri. Lesions of either of these regions give rise to *Sensory Aphasia*, which means a loss of speech due to inability to revive in memory the articulate sounds which serve as verbal symbols, or the graphic signs which serve as visual symbols for language.

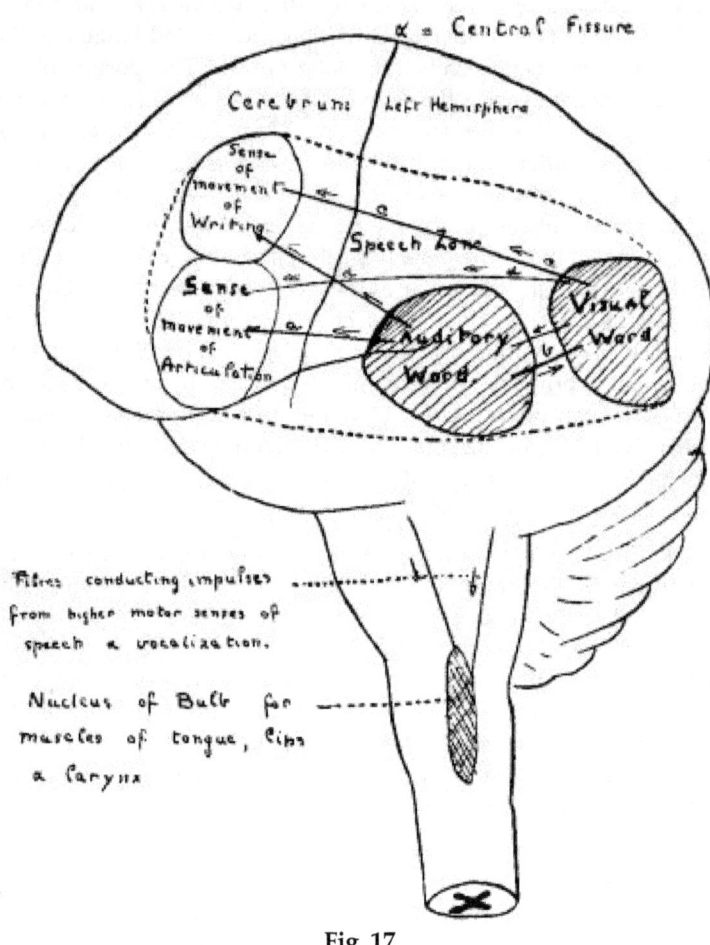

Fig. 17

FIG. 17.—Diagram to illustrate the Speech Zone of the left hemisphere (Bastian). This scheme is used to explain the mechanism of speech, but probably the centres are not precisely limited, as shown in the diagram; it serves, however, to explain disorders of speech. Destruction of the brain substance in front of the central fissure gives rise to what is termed Motor Aphasia and Motor Agraphia, because the patient no longer recalls the images of the movements necessary for expressing himself in articulate speech or by writing. Destructive lesions behind the central fissure may damage the portion of the brain connected with the mental perception of the sounds of articulate language, or the portion of the brain connected with the mental perception of language in the form of printed or written words—Sensory Aphasia; the former entails inability to speak, the latter inability to read.

This speech zone acts as a whole, and many disorders of speech may arise from destructive lesions within its limits. It has a special arterial supply, viz. the middle cerebral, which divides into two main branches—an anterior, which supplies the motor portion, and a posterior, which supplies the posterior sensory portion. The anterior divides into two branches and the posterior into three branches, consequently various limited portions of the speech zone may be deprived of blood supply by blocking of one of these branches. The speech zone of the left hemisphere directly controls the centres in the medulla oblongata that preside over articulation and phonation; innervation currents are represented by the arrows coming from the higher to the lower centres.

These several cortical regions are connected by systems of subcortical fibres to two regions in front of the ascending frontal convolution (*vide* fig. 17), called respectively the "glosso-kinfsthetic" (sense of movement of tongue) and the "cheiro-kinfsthetic" (sense of movement of hand) centres. Now a person may become hemiplegic and lose his speech owing either to the blood clotting in a diseased vessel, or to detachment of a small clot from the heart, which, swept into the circulation, may plug one of the arteries of the brain. The arteries branch and supply different regions, consequently a limited portion of the great brain may undergo destruction, giving rise to certain localising symptoms, according to the situation of the area which has been deprived of its blood supply. Upon the death of the patient, a correlation of the symptoms observed during life and the loss of brain substance found at the *post-mortem* examination has enabled neurologists to associate certain parts of the brain surface with certain functions; but M. Marie very rightly says: None of the older observations by Broca and others can be accepted because they were not examined by methods which would reveal the extent of the damage; the only cases which should be considered as scientifically reliable are those in which a careful examination by sections and microscopic investigation have determined how far subcortical structures and systems of fibres uniting various parts of the cortex in the speech zone have been damaged. Marie maintains that the speech zone cannot be separated into these several centres, and that destruction of Broca's convolution does not cause loss of speech (*vide* figs. 16, 17). There are at present two camps—those who maintain the older views of precise cortical centres, and those who follow Marie and insist upon a revision.

Herbert Spencer says that "our intellectual operations are indeed mostly confined to the auditory feelings as integrated into words and the visual feelings as integrated into ideas of objects, their relations and their motions."

Stricker by introspection and concentration of attention upon his own speech-production came to the conclusion that the primary revival of words was by the feeling of movements of the muscles of articulation; but there is a fallacy here, for the more the attention is concentrated upon any mental process the more is the expressive side brought into prominence in consciousness. This can be explained by the fact that there is in consequence of attention an increased outflow of innervation currents to special lower executive centres, thence to the muscles, but every change of tension in the speech muscles is followed by reciprocal incoming impressions appertaining to the sense and feeling of the movement. The more intense the sense of movement, the greater will be the effect upon consciousness. In fact, a person who reads and thinks by articulating the words, does so because experience has taught him that he can concentrate his attention more perfectly; therefore his memory or understanding of the subject read or thought of will be increased. Very many people think and commit to memory by this method of concentrating attention; they probably do not belong to the quick, perceptive, imaginative class, but rather to those who have power of application and who have educated their minds by close voluntary attention. Galton found a large proportion of the Fellows of the Royal Society were of this motor type. But the fact that certain individuals make use of this faculty more than others does not destroy the arguments in favour of the primary revival of words in the great majority of persons by a subconscious process in the auditory centre, which is followed immediately by correlated revival of sensori-motor images. Although the sensori-motor images of speech can be revived, it is almost impossible without moving the hand to revive kinfsthetic impressions concerned in writing a word. Both Ballet and Stricker admit this fact, and it tends to prove that the sense of hearing is the primary incitation to speech.

Charcot in reference to the interpretation of speech defects divided persons into four classes—auditives, visuals, motors, and indifferents. There are really no separate classes, but only different kinds of word-memory in different degrees of excellence as regards the first three; and as regards the fourth there is no one kind of memory developed to a preponderating degree. Bastian doubts the second class, but does not deny that the visual type may exist; for Galton has undoubtedly shown that visual memory and power of recall of visual word images varies immensely in different individuals, and it is unquestionable that certain individuals possess the visualising faculty to an extraordinary degree; some few, moreover, can see mentally every word that is uttered; they give their attention to the visual symbolic equivalent and not to the auditory. Such persons may, as Ribot supposes, habitually think and represent objects by visual typographic images. Lord Macaulay and Sir James Paget were notable possessors of this visualising faculty. The former is said to have been able to read a column of "The Times" and repeat it *verbatim*; the latter could deliver his lectures *verbatim* as he had written them. Both saw mentally the print or MS. in front of them.

Nevertheless it is a question of degree how much motor images enter into silent thought and into the primary revival of words in different individuals. Mach in "Analysis of Sensations" says: "It is true that in my own case words (of which I think) reverberate loudly in my ear. Moreover, I have no doubt that thoughts may be directly excited by the ringing of a house-bell, by the whistle of a locomotive, etc., that small children and even dogs understand words which they cannot repeat. Nevertheless I have been convinced by Stricker that the ordinary and most familiar, though not the only possible way, by which speech is comprehended is really *motor* and that we should be badly off if we were without it. I can cite corroborations of this view from my own experience. I frequently see strangers who are endeavouring to follow my remarks slightly moving their lips."

THE PRIMARY SITE OF REVIVAL OF WORDS IN SILENT THOUGHT

Since destructive lesions of the speech zone of the left hemisphere in right-handed persons leads to inability to revive the memory pictures of the sounds of words as heard in ordinary speech, the revival of visual impressions as seen in printed or written characters, and of the kinfsthetic (sense of movement) impressions concerned with the alterations of the minute tensions of the muscle structures employed in the articulation of words, it must be presumed that the left hemisphere in right-handed persons is dominant in speech and silent thought; it may even dominate the use of the left hand for many movements. But does not the right hemisphere take a part? Yes; and I will give my reasons later for supposing that the whole brain is in action. During the voluntary recall of words in speech and thought by virtue of the intimate association tracts connecting the grey matter of the whole speech zone, it is not a single part of this zone which is in action, but the whole of it; and when we assign to definite parts of the speech zone different functions in connection with language, we really refer to areas in which the process is most active or is primarily initiated, for the whole brain is in action just as it is in the recognition of an object which we see, hear, feel, or move. What really comes before us is contributed more by the mind itself than by the present object.

There is, however, a direct functional association between the auditory and glosso-kinfsthetic (sense of movement of the tongue) centres on the one hand and the visual and cheiro-kinfsthetic (sense of movement of the hand) on the other. No less intimate must be the connection between the auditory word-centre and the visual word-centre; they must necessarily be called into association actively in successive units of time, as in reading aloud or writing from dictation. Educated deaf mutes think with revived visual symbols either of lips or fingers. Words are to a great extent symbols whereby we carry on thought, and thinking becomes more elaborate and complex as we rise in the scale of civilisation, because more and more are verbal symbols instituted for concrete visual images.

In which portion of the brain are words primarily and principally revived during the process of thinking? I have already alluded to the views of Stricker and those who follow him, viz. that words are the revived images of the feelings of the sense of movement, caused by the alteration in the tension of the muscles of articulation occurring during speech, with or without phonation. There is another which I think the correct view, that words are revived in thought primarily as auditory images, so that the sense of hearing is essential for articulation as well as phonation; the two operations of the vocal organ as an instrument of the mind being inseparable. The arguments in favour of this are:—

1. The part of the brain concerned with the sense of hearing develops earlier and the nerve fibres found in this situation are myelinated[3] at an earlier period of development of the brain than the portion connected with the sense of movement of the muscles of articulation.

> [Footnote 3: The covering of the fibres by a sheath of phosphoretted fat serving to insulate the conductile portion of the nerve is an indication that the fibre has commenced to function as a conductor of nervous impulses.]

2. As a rule, the child's first ideas of language come through the sense of hearing; articulate speech is next evolved, in fact the child speaks only that which it has heard; it learns first to repeat the names of persons and objects with which it comes into relation, associating visual images with auditory symbols.

An example of this was communicated by Darwin to Romanes. One of his children who was just beginning to speak, called a duck a "quack." By an appreciation of the resemblance of qualities it next extended the term "quack" to denote all birds and insects on the one hand, and all fluid objects on the other. Lastly, by a still more delicate appreciation of resemblance the child called all coins "quack" because on the back of a French sou it had seen the representation of an eagle (Romanes' "Mental Evolution in Man," p. 183). Later on, children who have been educated acquire a knowledge of the application of visual symbols, and how to represent them by drawing and writing, and associate them with persons and objects.

3. There is more definiteness of impression and readiness of recall for auditory than for articulatory motor sense feelings.

4. After the acquirement of speech by the child, auditory feelings are still necessary for articulate speech processes; for if it were not so, how could we explain the fact that a child up to the fifth or sixth year in full possession of speech will become dumb if it loses the sense of hearing from middle-ear disease, unless it be educated later by lip language.

5. Cases have been recorded of bilateral lesion of the auditory centre of the brain producing loss of hearing and loss of speech, the motor centres being unaffected. This is called Wernicke's sensory aphasia. The following case occurring in my own practice is probably the most complete instance recorded.

CASE OF DEAFNESS ARISING FROM DE-STRUCTION OF THE AUDITORY CENTRES IN THE BRAIN CAUSING LOSS OF SPEECH

A woman at the age of twenty suddenly became unconscious and remained so for three hours; on recovery of consciousness it was found she could not speak; this condition remained for a fortnight; speech gradually returned, although it was impaired for a month or more. She married, but soon after marriage she suddenly lost her hearing completely, remaining permanently stone deaf; and although she could understand anything of a simple character when written, and was able imperfectly to copy sentences, she was unable to speak. Once, however, under great emotional excitement, while I was examining her by written questions, she uttered, "Is that." But she was never heard to speak again during the subsequent five years that she lived. The utterance of those two words, however, showed that the loss of speech was not due to a defect of the physiological mechanism of the vocal instrument of speech, nor to the motor centres in the brain that preside over its movements in the production of articulate speech. She recognised pictures and expressed satisfaction or dissatisfaction when correct or incorrect names were written beneath the pictures; moreover, in many ways, by gestures, facial expression, and curious noises of a high-pitched, musical, whining character, showed that she was not markedly deficient in intelligence. Although in an asylum and partially paralysed, she was not really insane in the proper sense, but incapable of taking care of herself. When other patients were getting into mischief this patient would give a warning to the attendants by the utterance of inarticulate sounds, showing that she was able to comprehend what was taking place around and reason thereon, indicating thereby that although stone deaf and dumb, it was probable that she possessed the power of silent thought. I observed that during emotional excitement the pitch of the sounds she uttered increased markedly with the increase of excitement. After having been discharged from Claybury Asylum she was sent to Colney Hatch Asylum. Upon one of my visits to that institution I learnt that she had been admitted, and upon my entering the ward, although more than a year had elapsed since I last saw her, she immediately and from afar recognised me; and by facial expression, gesture, and the utterance of inarticulate sounds showed her great pleasure and satisfaction in seeing one who had taken a great interest in her case. This poor woman must have felt some satisfaction in knowing that someone had interpreted her mental condition, for of course, her husband and friends did not understand why she could not speak. I may mention that the first attack of loss of speech was attributed to hysteria.

This woman died of tuberculosis seven years after the second attack, and examination of the brain *post-mortem* revealed the cause of the deafness. There was destruction of the centre of hearing in both hemispheres (*vide* fig. 17), caused by blocking of an artery supplying in each hemisphere that particular region with blood. The cause of the blocking of the two arteries was discovered, for little warty vegetations were found on the mitral valve of the left side of the heart. I interpreted the two attacks thus: one of these warty vegetations had become detached, and escaping into the arterial circulation, entered the left carotid artery and eventually stuck in the posterior branch of the middle cerebral artery, causing a temporary loss of word memory, consequently a disturbance of the whole speech zone of the left hemisphere. This would account for the deafness to spoken language and loss of speech for a fortnight, with impairment for more than a month, following the first attack. But both ears are represented in each half of the brain; that is to say, sound vibrations entering either ear, although they produce vibrations only in one auditory nerve, nevertheless proceed subsequently to both auditory centres. The path most open, however, for transmission is to the opposite hemisphere; thus the right hemisphere receives most vibrations from the left ear and *vice versa*. Consequently the auditory centre in the right hemisphere was able very soon to take on the function of associating verbal sounds with the sense of movement of articulate speech and recovery took place. *But*, when by a second attack the corresponding vessel of the opposite half of the brain was blocked the terminal avenues, and the central stations for the reception of the particular modes of motion associated with sound vibration of all kinds were destroyed *in toto*; and the patient became stone deaf. It would have been extremely interesting to have seen whether, having lost that portion of the brain which constitutes the primary incitation of speech, this patient could have been taught lip language.

There is no doubt that persons who become deaf from destruction of the peripheral sense organ late in life do not lose the power of speech, and children who are stone deaf from ear disease and dumb in consequence can be trained to learn to speak by watching and imitating the movements of articulation. Helen Keller indeed, although blind, was able to learn to speak by the education of the tactile motor sense. By placing the hand on the vocal instrument she appreciated by the tactile motor sense the movements associated with phonation and articulation. The tactile motor sense by education replaced in her the auditory and visual senses. The following physiological experiment throws light on this subject. A dog that had been deprived of sight by removal of the eyes when it was a puppy found its way about as well as a normal dog; but an animal made blind by removal of the occipital lobes of the brain was quite stupid and had great difficulty in finding its way about. Helen Keller's brain, as shown by her accomplishments in later life, was a remarkable one; not long after birth she became deaf and blind, consequently there was practically only one avenue of intelligence left open for the education of that brain, viz. the tactile kinfsthetic. But the tactile motor sense is the active sense that waits upon and contributes to every other sense. The hand is the instrument of the mind and the agent of the will; consequently the tactile motor sense is intimately associated in its structural representation in the brain with every other sense. This avenue being open in Helen Keller, was used by her teacher to the greatest possible advantage, and all the innate potentialities of a brain naturally endowed with remarkable intellectual powers were fully developed, and those cortical structures which normally serve as the terminal stations (*vide* fig. 16) for the reception and analysis of light and sound vibrations were utilised to the full by Helen Keller by means of association tracts connecting them with the tactile motor central stations. The brain acts as a whole in even the simplest mental processes by virtue of the fact that the so-called functional centres in the brain are not isolated fields of consciousness, but are inextricably associated one with another by association fibres.

THE PRIMARY REVIVAL OF SOME SENSATIONS IN THE BRAIN

I have on page 77 referred to Stricker's views on the primary revival of words in the sense of movement of the lips and tongue. Mach ("Analysis of the Sensations") says: "The supposition that the processes in the larynx during singing have had something to do with the formation of the tonal series I noticed in one of my earlier publications, but did not find it tenable. Singing is connected in too extrinsic and accidental a manner with hearing to bear out such an hypothesis. I can hear and imagine tones far beyond the range of my own voice. In listening to an orchestral performance with all the parts, or in having an hallucination of such a performance, it is impossible for me to think that my understanding of this broad and complicated sound-fabric has been effected by my *one* larynx, which is, moreover, no very practised singer. I consider the sensations which in listening to singing are doubtless occasionally noticed in the larynx a matter of subsidiary importance, like the pictures of the keys touched which when I was more in practice sprang up immediately into my imagination on hearing a performance on the piano or organ. When I imagine music, I always distinctly hear the notes. Music can no more come into being merely through the motor sensations accompanying musical performances, than a deaf man can hear by watching the movements of players. I cannot therefore agree with Stricker on this point" (comp. Stricker, "Du langage et de la musique," Paris, 1885).

Of the motor type myself and having a fairly good untrained ear for music, I find that to memorise a melody, whether played by an instrument or by an orchestra, I must either try to sing or hum that melody in order to fix it in my memory. Every time I do this, association processes are being set up in the brain between the auditory centres and the centres of phonation; and when I try to revive in my silent thoughts the melody again, I do so best by humming aloud a few bars of the melody to start the revival and then continuing the revival by maintaining the resonator in the position of humming the tune, viz. with closed lips, so that the sound waves can only escape through the nose; under such circumstances the only definite conscious muscular sensation I have is from the effect of closure of the lips; the sensations from the larynx are either non-existent or quite ill-defined, although I hear mentally the tonal sensations of the melody. No doubt by closing the lips in silent humming I am in some way concentrating attention to the sensori-motor sphere of phonation and articulation, and by reactive association with the auditory sphere reinforcing the tonal sensations in the mind. The vocal cords (ligaments) themselves contain very few nerve fibres; those that are seen in the deeper structures of the cords and adjacent parts mainly proceed to the mucous glands. This fact, which I have ascertained by numerous careful examinations, is in accordance with the fact that there are no conscious kinfsthetic impressions of alterations of position and tension of the vocal cords. A comparative microscopic examination of the tip of the tongue and the lips shows a remarkable difference, for these structures are beset with innumerable sensory nerves, whereby every slightest alteration of tension and minute variations in degrees of pressure of the covering skin is associated with messages thereon to the brain. The sense of movement in articulate speech is therefore explained by this fact. There is every reason then to believe that auditory tonal images are the sole primary and essential guides to the minute alterations of tension in the muscles of the larynx necessary for the production of corresponding vocal sounds. By humming a tune we concentrate our attention and thereby limit the activity of neural processes to systems and communities of neurones employed for the perception of tonal images and their activation in motor processes; and this helps to fix the tune in the memory.

PSYCHIC MECHANISM OF THE VOICE

A musical speaking voice denotes generally a good singing voice, and it must be remembered that articulation cannot be separated from phonation in the psychic mechanism. In speaking, we are unconscious of the breath necessary for the production of the voice. Not so, however, in effective singing, the management of the breathing being of fundamental importance; and it is no exaggeration to say that only the individual who knows how to breathe knows how to sing effectually. A musical ear and sense of rhythm are innate in some individuals; in others they are not innate and can only be acquired to a variable degree of perfection by persevering efforts and practice. The most intelligent persons may never be able to sing in tune, or even time; the latter (sense of rhythm) is much more easily acquired by practice than the former (correct intonation). This is easily intelligible, for rhythmical movement appertains also to speech and other acts of human beings, e.g. walking, dancing, running, swimming, etc.; moreover, rhythmical periodicity characterises the beat of the heart and respiration.

But how does a trained singer learn to sing a song or to take part in an opera? He has to study the performances of two parts for the vocal instrument—the part written by the composer and the part written by the poet or dramatist—and in order to present an artistic rendering, the intellectual and emotional characters of each part must be blended in harmonious combination. A singer will first read the words and understand their meaning, then memorise them, so that the whole attention subsequently may be given to applying the musical part to them and employing with proper phrasing, which means more than knowing when to breathe; it means imparting expression and feeling. A clever actor or orator can, if he possess a high degree of intelligence and a fairly artistic temperament, so modulate his voice as to convey to his audience the passions and emotions while feeling none of them himself; so many great singers who are possessed of a good musical ear, a good memory, and natural intelligence, although lacking in supreme artistic temperament and conspicuous musical ability, are nevertheless able to interpret by intonation and articulation the passions and emotions which the composer has expressed in his music and the poet or dramatist in his words. The intelligent artist possessed of the musical ear, the sense of rhythm, and a well-formed vocal organ accomplishes this by the conscious control and management of his breathing muscles and the muscles of articulation, which by education and imitation he has brought under complete control of the will. With him visual symbols of musical notes are associated with the visual symbols of words in the mind, and the visual symbols whether of the words or of the musical notes will serve to revive in memory the sound of the one or the other, or of both. But he produces that sound by alteration of tension in co-ordinated groups of muscles necessary for vocalisation, viz. the muscles of phonation in the larynx, the muscles of articulation in the tongue, lips, jaw, and palate, and the muscles of costal respiration. *The mind* of the orator, actor, and dramatic singer exercises a profound influence upon the respiratory system of nerves, and thereby produces the necessary variations in the force, continuance, and volume of air required for vocal expression.

Sir Charles Bell, who discovered the respiratory system of nerves, pointed out how the lungs, from being in the lower animals merely the means of oxygenating the blood, become utilised in the act of expelling air from the body for the production of audible sounds — the elements of human voice and speech. Likewise he drew attention to the influence which powerful emotions exercise upon the organ of respiration, including the countenance, e.g. the dilated nostrils in anger. Again, "when the voice suffers interruption and falters, and the face, neck, and chest are animated by strong passion working from within the breast, language exerts its most commanding influence."

In hemiplegia or paralysis of one half of the body, there is a difference between the two sides for ordinary automatic unconscious diaphragmatic breathing and voluntary or costal breathing. Thus in ordinary breathing the movements are increased on the paralysed side, especially in the upper part of the chest, while in voluntary breathing they are increased on the sound side. Hughlings Jackson suggested the following theory to explain these facts: "*Ordinary breathing* is an automatic act governed by the respiratory centre in the medulla. The respiratory centre is double, each side being controlled or inhibited by higher centres on the opposite side of the brain. Voluntary costal breathing, such as is employed in singing, is of cerebral origin, and controlled by centres on the opposite side of the brain, the impulses being sent down to the respective centres for the associated movements of the muscles of articulation, phonation, and breathing, in the same way as they are sent to the centres for the movements of the arm or leg. With voluntary breathing the respiratory centre in the medulla has nothing to do. It is in fact out of gear or inhibited for the time being, so that the impulses from the brain pass by or evade it. There are thus two sets of respiratory nerve fibres passing from the brain — the one inhibiting or controlling to the opposite half of the respiratory centre in the medulla; the other direct, evading the respiratory centre and running the same course to the spinal centres for the respiratory movements as the ordinary motor fibres do to the centres for other movements. Both sets would be affected by the lesion (or damage) which produced the hemiplegia. The inhibitory fibres being damaged, the opposite half of the respiratory centre would be under diminished control and therefore the movements of ordinary breathing on the paralysed side would be exaggerated. The damage to the direct fibres would prevent the passage of voluntary stimuli to the groups of respiratory muscles (as it would do to the rest of the muscles of the paralysed side), and thus the voluntary movement of respiration would be diminished — diminished only and not completely abolished as in the limbs; because according to the theory of Broadbent, in the case of such closely associated bilateral movements the lower nervous respiratory centres of both sides would be activated from either side of the brain." This certainly applies also to the muscles of phonation, but not to the principal muscles of articulation, viz. the tongue and lips. It is not exactly known what part of the cerebral cortex controls the associated movements necessary for voluntary costal (rib) respiration in singing; probably it is localised in the frontal lobe in front of that part, stimulation of which gives rise to trunk movements (*vide* fig. 16). Whatever its situation, it must be connected by association fibres with the centres of phonation and articulation.

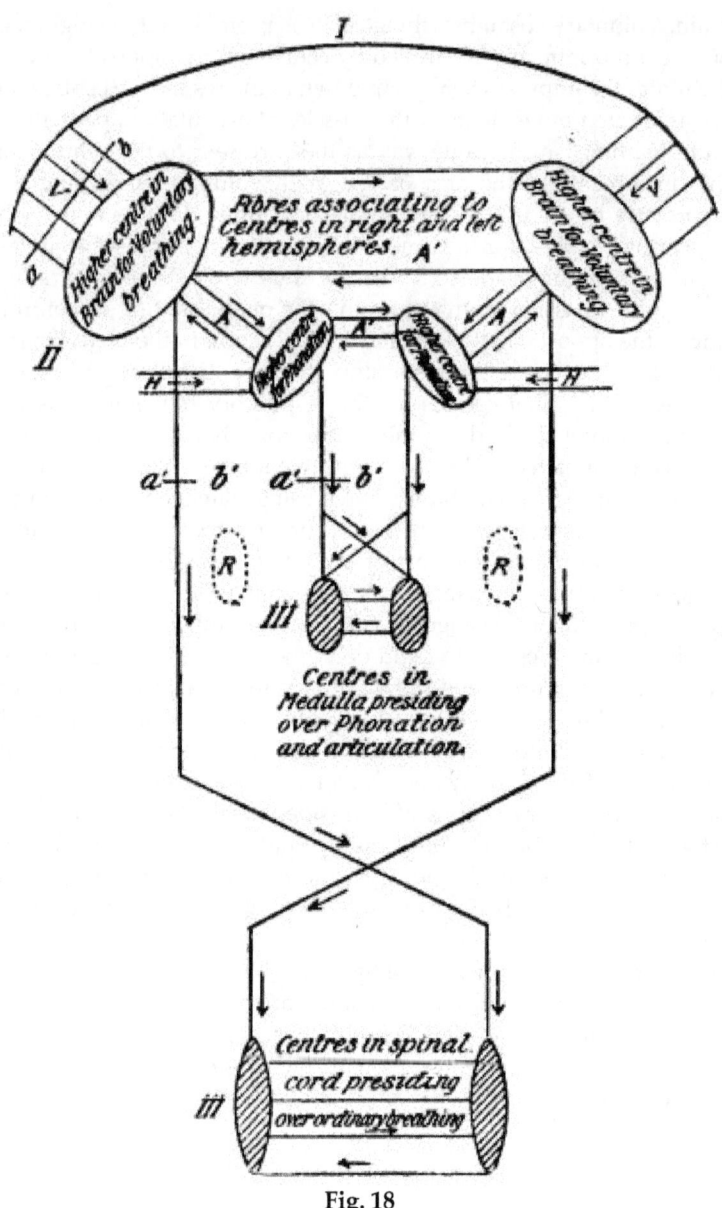

Fig. 18

FIG. 18.—The accompanying diagram is an attempt to explain the course of innervation currents in phonation.

1. Represents the whole brain sending voluntary impulses V to the regions of the brain presiding over the mechanisms of voluntary breathing and phonation. These two regions are associated in their action by fibres of association A; moreover, the corresponding centres in the two halves of the brain are unified in their action by association fibres A' in the great bridge connecting the two hemispheres (Corpus Callosum). On each side of the centre for phonation are represented association fibres H which come from the centre of hearing; these fibres convey the guiding mental images of sounds and determine exactly the liberation of innervation currents from the centre of phonation to the lower centres by which the required alterations in tension of the laryngeal muscles for the production of the corresponding sounds are effected. Arrows are represented passing from the centre of phonation to the lower centres in the medulla which preside over the muscles of the jaw, tongue, lips, and larynx. Arrows indicate also the passage of innervation currents from the centres in the brain which preside over voluntary breathing. It will be observed that the innervation currents which proceed from the brain pass over to the opposite side of the spinal cord and are not represented as coming into relation with the respiratory centre R. This centre, as we have seen, acts automatically, and exercises especially its influence upon the diaphragm, which is less under the control of the will than the elevators of the ribs and the abdominal muscles.

The diagram also indicates why these actions of voluntary breathing and phonation can be initiated in either hemisphere; it is because they are always bilaterally associated in their action; consequently both the higher centres in the brain and the lower centres in the medulla oblongata and spinal cord are united by bridges of association fibres, the result being that even if there is a destruction of the brain at *a-b*, still the mind and will can act through both centres, although not so efficiently. Likewise, if there is a destruction of the fibres proceeding from the brain centres to the lower medullary and spinal centres, the will is still able to act upon the muscles of phonation and breathing of both sides of the body because of the intimate connection of the lower medullary and spinal centres by association fibres.

Experiments on animals and observations on human beings show that the centres presiding over the muscles of the larynx are situated one in each hemisphere, at the lower end of the ascending frontal convolution in close association with that of the tongue, lips, and jaw. This is as we should expect, for they form a part of the whole cerebral mechanism which presides over the voice in speech and song. But because the muscles of the tongue, the lower face muscles, and even the muscles of the jaw do not necessarily and always work synchronously and similarly on the two sides, there is more independence in their representation in the cerebral cortex. Consequently a destruction of this region of the brain or the fibres which proceed from it to the lower executive bulbar and spinal centres is followed by paralysis of the muscles of the opposite side. Likewise stimulation with an interrupted electric current applied to this region of the brain in monkeys by suitable electrodes produces movements of the muscles of the lips, tongue, and jaw of the opposite side only. Not so, however, stimulation of the region which presides over the movements of the muscles of the larynx, for then *both* vocal cords are drawn together and made tense as in phonation. It is therefore not surprising if removal or destruction of this portion of the brain *on one side* does not produce paralysis of the muscles of phonation, which, always bilaterally associated in their actions, are represented as a bilateral group in both halves of the brain. These centres may be regarded as a part of the physiological mechanism, but the brain acts as a whole in the psychic mechanism of speech and song. From these facts it appears that there is: (1) An automatic mechanism for respiration and elemental phonation (the cry) in the medulla oblongata which can act independently of the higher centres in the brain and even without them (*vide* p. 18). (2) A cerebral conscious voluntary mechanism which controls phonation either alone or associated with articulation. The opening of the glottis by contraction of the abductor (posterior ring-pyramid muscles) is especially associated with descent of the diaphragm in inspiration in ordinary breathing; whereas the voluntary breathing in singing is associated with contraction of the adductor and tensor muscles of the vocal cords.

A perfect psychic mechanism is as necessary as the physiological mechanism for the production of perfect vocalisation, especially for dramatic singing. A person, on the one hand, may be endowed with a grand vocal organ, but be a failure as a singer on account of incorrect intonation, of uncertain rhythm or imperfect diction; on the other hand, a person only endowed with a comparatively poor vocal instrument, but knowing how to use it to the best advantage, is able to charm his audience; incapable of vigorous sound production, he makes up for lack of power by correct phrasing and emotional expression. We see then that the combination of a perfect physiological and psychological mechanism is essential for successful dramatic singing, the chief attributes of which are: (1) Control of the breath, adequate volume, sustaining power, equality in the force of expulsion of air to avoid an unpleasant vibrato, and capability of producing and sustaining loud or soft tones throughout the register. (2) Compass or range of voice of not less than two octaves with adequate control by mental perception of the sounds of the necessary variation in tension of the laryngeal muscles for correct intonation. (3) Rich quality or timbre, due partly to the construction of the resonator, but in great measure to its proper use under the control of the will. Something is lacking in a performance, however perfect the vocalisation as regards intonation and quality, if it fails to arouse enthusiasm or to stir up the feelings of an audience by the expression of passion or sentiment through the mentality of the singer.

The general public are becoming educated in music and are beginning to realise that shouting two or three high-pitched chest notes does not constitute dramatic singing—"a short *beau moment* does not compensate for a *mauvais quart d'heure.*" It would be hard to describe or define the qualities that make a voice appeal to the multitude. Different singers with a similar timbre of voice and register may sing the same song correctly in time, rhythm, and phrasing, and yet only one of them may produce that sympathetic quality necessary to awaken not only the intellectual but the affective side of the mind of the hearers. Undoubtedly the effects produced upon the mind by dramatic song largely depend upon circumstances and surroundings, also upon the association of ideas. Thus I was never more stirred emotionally by the human voice than upon hearing a mad Frenchman sing at my request the Marseillaise. Previously, when talking to him his eyes had lacked lustre and his physiognomy was expressionless; but when this broad-chested, six foot, burly, black-bearded maniac rolled out in a magnificent full-chested baritone voice the song that has stirred the emotions and passions of millions to their deepest depth, and aroused in some hope, in others despair, as he made the building ring with "Aux armes, citoyens, formez vos bataillons" I felt an emotional thrill down the spine and a gulp in the throat, while the heart and respirations for an instant stayed in their rhythmical course. Not only was I stirred by the effect of the sounds heard, but by the change in the personality of the singer. It awakened in my mind the scenes in the French Revolution so vividly described by Carlyle. The man's facial expression and whole personality suddenly appeared changed; he planted his foot firmly forward on the ground, striking the attitude of a man carrying a musket, a flag, or a pike; his eyes gleamed with fire and the lack-lustre expression had changed to one of delirious excitement. A pike in his hand and a red cap on his head would have completed the picture of a *sans culotte*. Dramatic song therefore that does not evoke an emotional response is *vox et prſterea nihil*.

www.ingramcontent.com/pod-product-compliance
Lightning Source LLC
Chambersburg PA
CBHW070310230526
45470CB00002B/807